Theatre, Performance and Cognition

Languages, Bodies and Ecologies

Edited by
Rhonda Blair and Amy Cook

Bloomsbury Methuen Drama
An imprint of Bloomsbury Publishing Plc

BLOOMSBURY

LONDON · OXFORD · NEW YORK · NEW DELHI · SYDNEY

Bloomsbury Methuen Drama
An imprint of Bloomsbury Publishing Plc

Imprint previously known as Methuen Drama

50 Bedford Square   1385 Broadway
London   New York
WC1B 3DP   NY 10018
UK   USA

www.bloomsbury.com

**BLOOMSBURY, METHUEN DRAMA and the Diana logo are trademarks of
Bloomsbury Publishing Plc**

First published 2016

© Rhonda Blair and Amy Cook and contributors, 2016

British Library Cataloguing-in-Publication Data
A catalogue record for this book is available from the British Library.

ISBN: HB: 978-1-4725-9179-1
PB: 978-1-4725-9178-4
ePDF: 978-1-4725-9181-4
ePub: 978-1-4725-9180-7

Library of Congress Cataloging-in-Publication Data
A catalog record for this book is available from the Library of Congress.

Typeset by Deanta Global Publishing Services, Chennai, India
Printed and bound in Great Britain

Theatre, Performance and Cognition

WITHDRAWN

Performance and Science: Interdisciplinary Dialogues explores the interactions between science and performance, providing readers with a unique guide to current practices and research in this fast-expanding field. Through shared themes and case studies, the series offers rigorous vocabularies and methods for empirical studies of performance, with each volume involving collaboration between performance scholars, practitioners and scientists. The series encompasses the modalities of performance to include drama, dance and music.

Series Editors

John Lutterbie
Chair of the Departments of Art and of Theatre Arts at Stony Brook University, USA

Nicola Shaughnessy
Professor of Performance at the University of Kent, UK

In the same series

Affective Performance and Cognitive Science
edited by Nicola Shaughnessy
ISBN 978-1-4081-8398-4

Performance and the Medical Body
edited by Alex Mermikides and Gianna Bouchard
ISBN 978-1-4725-7078-9

Theatre and Cognitive Neuroscience
edited by Clelia Falletti, Gabriele Sofia and Victor Jacono
ISBN 978-1-4725-8478-6

Contents

List of Figures

List of Contributors

Rhonda Blair is professor of theatre in the Division of Theatre, Southern Methodist University. She is the author of *The Actor, Image, and Action: Acting and Cognitive Neuroscience* (2008) and essays in numerous journals and edited volumes. She directs and performs, and is past president of the American Society for Theatre Research (2009–12).

Amy Cook is an associate professor in the Departments of English and Theatre Arts at Stony Brook University. She is the author of *Shakespearean Neuroplay* (2010) and has published essays in several journals and edited volumes, including *The Oxford Handbook of Dance and Theatre* (2015) and the forthcoming *Shakespeare and Consciousness*. You can find more about her at: https://sbsuny.academia.edu/AmyCook.

Barbara Dancygier is a professor in the Department of English, University of British Columbia, Canada. Her interdisciplinary work combines interests in language, cognition and multimodal forms of communication. Her publications in cognitive linguistics, cognitive narratology and cognitive poetics bring together the study of language and work on visual artefacts, materiality, embodiment and performance. She has published three monographs and three edited volumes (see http://faculty.arts.ubc.ca/bdancygier/index.html).

Shaun Gallagher is the Lillian and Morrie Moss Professor of Excellence in Philosophy at University of Memphis (United States) and Professorial Fellow, Faculty of Law, Humanities, and the Arts, University of Wollongong (Australia). He is also a Humboldt Foundation Anneliese Maier Research Fellow (2012–17).

Matt Hayler is a lecturer in post-1980 fiction at the University of Birmingham. His research explores the representations of technology and embodiment in all kinds of artwork and public discourse, drawing on cognitive science, (post)phenomenology and critical theory to consider the entanglement of user, tool and cultural milieu. His recent book, *Challenging the Phenomena of Technology*, uses the move from reading paper books to reading digital screens as a way in to considering the nature of technology more broadly. This work led to his representing the UK as a management committee member for the COST European E-READ research network. He is currently researching trans-humanism, the modification of the body with technology, and digital currencies.

Christopher J. Jackman teaches at Brock University, Canada, and received his doctorate from University of Toronto's Graduate Centre for Drama, Theatre and Performance Studies. He is founder of the youth theatre programme Lock and Keynote Productions, and director of the site-specific devising ensemble, Open Corps Theatre. His upcoming publications include *War Horse: Performance, Cognition, and the Spectator* (Palgrave, with Dr Toby Malone). Christopher resides in Toronto with his wife Shari and daughter Susanna.

Sarah E. McCarroll is assistant professor of theatre at Georgia Southern University where she teaches Costume Design and Theatre History. She is also a faculty member of the Center for Irish Research and Teaching. Sarah has been a member of the costume department staff at the Utah Shakespeare Festival since 2003, where she has served in a variety of positions including First Hand and Draper. Her article 'The "Boy" Who Wouldn't Grow Up: Peter Pan and the Dangers of Eternal Youth' appears in *Theatre Symposium 23*.

Laura Seymour wrote her PhD thesis on cognitive theory and Shakespeare's plays at Birkbeck, University of London. She is currently working on turning her thesis into a monograph and on a discussion of stillness in *The Taming of the Shrew*, forthcoming in 2016 in *The Cognitive Humanities: Embodied Mind in Literature and Culture*, ed. Peter Garratt.

Catherine J. Stevens is a cognitive psychologist who investigates the psychological processes in creating, perceiving and performing music and dance. She is the author of more than 170 articles, book chapters, conference proceedings papers and an e-book on creativity and cognition in contemporary dance. She is also a professor in psychology and director of research and engagement in the MARCS Institute for Brain, Behaviour and Development at Western Sydney University (http://katestevens.weebly.com).

Vera Tobin is assistant professor of cognitive science at Case Western Reserve University. Her recent work has included publications on detective fiction, viewpoint in language, and presupposition, and she is currently at work on a book-length study of cognitive bias and the poetics of surprise. To find more about her, visit: https://case.academia.edu/VeraTobin.

Mark Turner is Institute Professor and professor of cognitive science at Case Western Reserve University. He is the founding director of the Cognitive Science Network (where you can find many of his papers) and co-director of the Red Hen Lab. His most recent book publications are *The Origin of Ideas: Blending, Creativity, and the Human Spark* (Oxford University Press, 2014), *Ten Lectures on Mind and Language* (Eminent Linguists Lecture Series. Beijing: FLTR Press, 2011) and two edited volumes, *The Artful Mind: Cognitive Science and the Riddle of Human Creativity*, from Oxford University Press, and *Meaning, Form, & Body*, edited with Fey Parrill and Vera Tobin, published by the Center for the Study of Language and Information. He has been a fellow of the Institute for Advanced Study, the John Simon Guggenheim Memorial Foundation, the National Humanities Center, the National Endowment for the Humanities and the Institute of Advanced Study of Durham University, among others. In 2015, the Alexander von Humboldt Foundation awarded him an Annaliese Maier Research Prize for work in multimodal communication.

Evelyn B. Tribble is professor and Donald Collie Chair of English at the University of Otago, Dunedin, New Zealand. She is the author of *Margins and Marginality: The Printed Page in Early Modern England* (Virginia, 1993); *Writing Material: Readings from Plato to the Digital Age* (with Anne Trubek, Longmans, 2003); *Cognitive Ecologies and the History of Remembering* (with Nicholas Keene, Palgrave, 2011); and *Cognition in the Globe: Attention and Memory in Shakespeare's Theatre* (Palgrave, 2011). She has also published scholarly articles in *Shakespeare Quarterly*; *Shakespeare*; *Shakespeare Survey*; *ELH*, *Shakespeare Studies*; and *Textual Practice*, among others.

Neal Utterback is assistant professor of theatre at Juniata College. His current research examines intersections of sports psychology, cognitive science and acting theory in an effort to better prepare young artists for the psychophysical demands of performance. Recently, he premiered *Citizen Cyborg*, adapted from James Hughes's seminal work on trans-humanism and democracy of the same name, at New York's Planet Connections Theatre Festivity.

Edward C. Warburton is professor of dance and associate dean of the arts at UC Santa Cruz. His research explores the cognitive processes and relational practices that enhance (or undermine) the doing, making and viewing of dance (find more about him at: people.ucsc.edu/tedw).

Acknowledgements

This work came out of several years of working with the American Society for Theatre Research's Working Group in Cognitive Science in Theatre, Dance, and Performance. Some of the usual suspects who deserve mention: John Lutterbie, Bruce McConachie, Rick Kemp, Naomi Rokotnitz, Stanton Garner, Diana Calderazzo, Nicola Shaughnessy, Pamela Decker, Vanille Roche-Fogli, Maiya Murphy, Sara Taylor, Teemu Paavolainen, David Bisaha, Stephen DiBenedetto, Eric Heaps, Vivian Appler, Laura Lodewyck, Collin Bjork, Slade Billew, Gabriele Sofia and many others.

The book came together thanks to key help from Kristin Leadbetter and Daniel Irving.

Rhonda Blair thanks Southern Methodist University's University Research Council and SMU's Dedman College Interdisciplinary Institute for financial assistance; Robert Howell for collaboration cochairing the Situated Self seminar and for valuable conversations about intersections of philosophy and cognitive science that have informed her thinking for this book. For his unending support, she would like to thank Bill Beach.

Amy Cook thanks Indiana University's College of Arts and Sciences for travel support and Stony Brook University's Center for Embodied Cognition, Creativity and Performance for intellectual travel. For emotional travel and support, she would like to thank Ken Weitzman.

Acknowledgements

Introduction

Rhonda Blair and Amy Cook

Stories told by bodies onstage. We have not stopped being interested in stories or those bodies, or in coming together to hear them. The language, bodies and stages change, but their fascination remains. Theatre and performance moves us, and making sense of how we are moved and what it takes to move another through performance requires a convergence of methods, evidence, lenses and insights. In the last twenty years or so, scholars in the arts and humanities have integrated findings from the cognitive sciences into their research. This book brings together scholarship in three areas in which scholars are using the sciences to inform their thinking about theatre and performance: language, bodies and ecologies.

Cognitive approaches have opened up new avenues of research, new questions and new answers. We cannot derive the answers we need from psychoanalysis or post-structuralism or post-colonialism – which is not the same as saying that all the insights from these must be jettisoned. We are interested in understanding how bodies tell stories onstage and why audiences continue to come and participate. We want to know more fully the relationship between language and cognition – how can we better probe the words characters say for the experience that language is meant to generate? Are words symptoms of a hidden truth, or do they reveal in ways we haven't been examining? We want to know how bodies onstage capture our attention – how can we improve our understanding of what happens during performance for the actor/dancer in order to adjust our training to reflect what it is that is actually required of them? What can we learn from the extraordinary cognitive feats performed by a dancer reproducing choreography night after night, town after town? We want to know how the performance event works as an ecosystem – what drives the system of performer and environment? To what degree are we staging the environments that have shaped the characters, rather than just the characters in their environments? These questions and others like them are being fruitfully engaged using research from the sciences.

As the chapters in this volume attest, there is no single methodology to this work. There are conversations happening across the disciplines on areas of significance for our study. There is, however, a general agreement about the science. None of the chapters in this book argue for a Cartesian self with a thinking brain and an animating soul; emotions are not viewed as distractions from cognition, rather they are part of it – the body is not separate from the brain. None of the chapters in this book think of language as generative based on rules and systems in the brain. In this introduction we provide the reader with a map of the cognitive science terrain, with particular focus on the research that has been the most influential on the essays herein. The cognitive science of today is part of a history of developments; we do not view the science as fixed and stable, but rather as building on previous theories, discoveries and paradigm shifts.

Before we speak of what today's science is, however, we would like to make very clear what we believe it is not. There are those who fear that the cognitive turn means that we will all need to do experiments following scientific methodologies, or even that empirical verification will be required as evidence in all situations. This is not true. Though some scholars are interested in collaborating with scientists on experiments, many are more interested in making connections between empirical work in one discipline and theoretical work in another, or in appropriating the science actually or metaphorically for applications in the studio. We understand – as do most scientists – the significant limitations on scientific experiments and are not suggesting that that work is superior. In order to conduct an experiment, one must reduce as much complexity (or 'noise') as possible; many times, this squeezes the very thing we care about out of the room. In order to conduct an experiment, one must necessarily control, contain or watch a phenomenon that might either alter while being watched or be invisible to our modes of looking. In addition to the many limitations on fMRI studies, for example, it would be impossible to scan the brains of an engaged audience in situ; unless we are willing to ignore the impact of the body of the spectator, the surrounding audience members and almost all other elements of the live event, an fMRI will not tell us what we are interested in. A cognitive approach to performance need not be empirical.

An approach that calls on the sciences is also not necessarily more right or valid than one that does not. It is useful in many cases if work in one field is in conversation with another – particularly when both are interested in some of the same questions. Psychologists, cognitive scientists, cognitive linguists and literary scholars are all interested, albeit in different ways, in how language works. They ask different questions, use different methods, probe different archives and present their findings in different ways, but are all guided by assumptions and hypotheses about the relationship between language and thinking. It seems only fitting that there be bridges across these divides for conversation and coherence around these assumptions. If language is not composed of a system of difference or symbols, connecting things in the world with ideas in the brain, if, in fact, it is more creative and generative than that, then the way literature and performance work to move us, inspire us, engage us, is going to be different than we thought.

Finally, this approach does not require us to essentialize. A cognitive analysis of the phenomenon of spectating – what it means to take in information in a theatrical setting – does universalize the experience insofar as the biologies of the agents are more similar than they are different; but *that* an Elizabethan man and a twenty-first-century Korean-American woman both understand an experience is different from thinking about *how* or *what* they understand. In the same way, a cognitive analysis of how any dancer or actor works is likely to find significant similarities in neural, cognitive and kinaesthetic operations, though these operations are understood and applied differently across cultures and 'ecologies'. The essays collected here may not explicitly address specific issues around culture, race, gender or class, but that is because the questions they are asking focus on the cognitive systems, not the cultural ones. More and more, though, we are learning how cognitive systems are inseparable from cultural systems – if cognition is as embodied, embedded, enacted and extended as many of us believe. In this case, there will be different questions to ask about the systems in place around race, gender and class.

What does it mean to say that cognition is embodied? Embodied cognition is not something that requires explicit physical movement or action; cognition is embodied when we add up receipts, get a glass of water for a thirsty friend or reach for the shampoo in the shower. To say that cognition is embodied is to say that what we have called 'thinking' requires the body and happens as it does because of the body we have. The problem is not with making our thinking or our performing more 'embodied' – thought is always embodied. The challenge is coming up with language to articulate what it has been all along: we have missed some of the nuances because our language looks for bodies and minds. The paradigm shift that keeps trying to arrive is about recategorizing what we are talking about when we talk about thinking, or meaning, or touching, or feeling, or embodiment or performance. This requires close reading and creativity and a facility with metaphor. All the money in the sciences cannot do what those of us in the arts and humanities can do: give us new ways to see and stage – and thus to understand and to use and to develop – the implications of embodied, embedded, enacted and extended minds.

Perhaps it is helpful to imagine what our 'mother's cognitive science' thought about thinking. Thinking happened in the brain. The brain solved problems based on input and it generated and distributed solutions. The brain's power could be swayed or disrupted by the occasional emotional or physical experience, but mostly it kept itself above that fray. The brain told the body to move and the body responded. The stream of chit-chat in the head was evidence of this thinking. The ability to take in stimuli and translate it from visual symbol to meaning (with associations and memories) and then formulate a response was the power of the brain. The brain saw leaves moving on the trees, deduced that there must be wind moving them, concluded that the season might be changing and planned to get the sweaters out of the basement. Reading required the brain to turn lines on a page into symbols that attached to real-world things or events through a set of rules around nouns, verbs and prepositions. Poetry was the fantastic work of the madman and artist: using words in ways that ran counter to their literal meaning to require the reader to venture into different territory to make sense of it. The phrases 'he looked out of the window' and 'he went out of his mind' were not considered metaphoric.

This seemed intuitively right for many years, and not just to our mothers. One obvious problem with this is that there is no 'little thinker' in the brain; we have no evidence of a central command control centre creating representations of what we see. Even if there were, it would not explain how the symbolic representation of 'chair' in our brains matches the external reality or use of the chair. It doesn't explain how we come to categorize the chair as something to sit *on* and the window as something that separates the inside from the out. The power of the container metaphor (the one that allows us to think of the chair, the window and our minds as being or belonging to a container) to structure experience – and language – is unexplained in traditional, literal, representational theories of cognition. How do we come to say that we see the wind moving the leaves when clearly we cannot see the wind? How do we shift attention from the chair to the wind? How are we able to make the chair (or the tree or the leaves) into a figure pulled out from the noisy ground of all the visual stimuli available to us from the chair in our office?

Perhaps, part of the confusion comes from the different kinds of things we say we 'know'. We can 'know' that a friend feels sad and we can 'know' that something is about to happen onstage and we can 'know' chair. The first involves emotions and empathy and a clear need to feel into the knowledge referred to – few would argue that thinking about another person's emotional state could be solely logical. The second is a dynamic cognition that folds past, present and future together. The third, though, is referential – a chair is a chair – and less obviously embodied, but understanding how we understand chair is one thing scientists have not been able to teach a robot to do. This is referred to as the symbol-grounding problem: How does a representation acquire meaning? Some now argue that a chair has no meaning independent of what we do to or with it. According to Alva Noë, 'We ought to reject the idea – widespread in both philosophy and science – that perception is a process *in the brain* whereby the perceptual system constructs an *internal representation* of the world. ... What perception is, however, is not a process in the brain, but a kind of skillful activity on the part of the animal as a whole' (2004: 2). Cognition is what happens when we sit on the chair.

The cognitive science(s)[1]

Cognitive science encompasses cognitive psychology, cognitive linguistics, neuroscience, neurolinguistics and cognitive anthropology, among other specializations. Some of the most important work came about because of the intersections of these (and other) specializations. Here, some attempts have been made lay out some of the complexities of the fields with a bit more clarity. Because it is easy to misappropriate the complex material produced by research in cognition, we must be mindful of being non-expert in the sciences and of the dangers of misunderstanding or mistranslating what we read. It is *crucial* to keep in mind that there is often disagreement among experts in the science disciplines, much as there is among performance studies scholars and practitioners.

What is sometimes called the second generation of cognitive science arose in the 1990s as cognitive scientists, neuroscientists, linguists and philosophers, among others, began to look at scientific findings in light of phenomenological and epistemological frameworks to begin to see mind as a manifestation of the body, inextricably linked to its environment – its ecology. This second phase is attempting to identify, among other things, how consciousness arises and its relationship to language, emotion and our interactions with the world, as an embodied process. Although we do not know all of the steps by which 'matter becomes imagination' (to use a phrase from neuroscientists Gerald Edelman and Giulio Tononi 2000), how elements of consciousness and behaviour relate to brain function is becoming clearer. Neuroscientist Antonio Damasio comes at the problem of consciousness through his somatic marker hypothesis, 'somatic marker' being a term for describing how body states become linked with conscious responses to or interpretations of them (Damasio 1994 and 1999: esp. 173–80). As with other views cited here, body, feeling and intellect are viewed as aspects of a single, if complex organic process. Particularly pertinent is Damasio's assertion that reason in the fullest sense grows out of and is permeated by emotion, and that emotion is consistently affected by reason and conscious cognition. Some neuroscientists theorize that who we are and how we function are based largely upon the development of specific neural patterns, or synaptic connections.

From one perspective, the self can be seen as our experience of what arises out of bodily dynamic neural, chemical and kinetic processes – only some, perhaps few, of which we consciously register as we interact with our environments. This 'ecological' view of the development and function of cognition is also evident in the work of Edelman and Tononi, who hold that higher brain functions, including consciousness, are conditioned by and require interactions with the world and other people, that is, mind is a result of reciprocal interaction between perceptual and proprioceptive experience, between external and internal environments, such that what happens in one influences what happens in the other. Only physical processes are needed to explain consciousness, and consciousness arose only because of a very specific evolution of the physical human body in response to environmental changes both in the natural world and in the organization of human communities. Rejecting the concept of the closed circuits of binary logic as the basis for understanding cognitive processes, embodied cognition is viewed as an open, non-linear system that is subject to perturbations from an array of different sources.

Cognition is embodied, that is, not separable from our physicality. Cognition is embedded, that is, it depends heavily on offloading cognitive work and taking advantage of affordances, or potentials, in the environment, and, as such, it is very much a result of 'ongoing agent-environment interaction' (Robbins and Aydede 2009: 10). Cognition is extended, that is, 'the boundaries of cognitive systems lie outside the envelope of individual organisms, encompassing features of the physical and social environment. ... In this view the mind leaks out into the world and cognitive activity is distributed across individuals and situations' (Robbins and Aydede 2009: 10).

This extended view of cognition is informed by mathematics' dynamical systems theory, which describes the operations of complex dynamic systems of interaction. In this view, because states of a cognitive system

> depend as much on changes in the external environment as on changes in the internal one, it becomes as important for cognitive modeling to track causal processes that cross the boundary of the individual organism as it is to track those that lie within that boundary. In short, insofar as the mind is a dynamical system, it is natural to think of it as extending not just into the body, but also into the world. (Robbins and Aydede 2009: 8)

To cite philosopher Evan Thompson, mind is 'an embodied dynamic system in the world', not a 'neural network in the head'; our world 'is not a prespecified, external realm, represented internally by its brain, but *a relational domain enacted or brought forth by that being's autonomous agency and mode of coupling with the environment*' (Thompson 2007: 10–11, and 13, italics ours). The relationship between the organism and the world is one of 'dynamic co-emergence', that is, 'cognition unfolds as the continuous coevolution of acting, perceiving, imagining, feeling, and thinking' (Thompson 2007: 60 and 43). The organism engages in *auto-poesis*, or self-making, within an *ecopoetic* situation; the self is not made without being made by and also making its environment (Thompson 2007: 119). As William Clancy writes,

> We cannot locate meaning in the text, life in the cell, the person in the body, knowledge in the brain, a memory in a neuron. Rather, these are all active, dynamic processes, existing only in interactive behaviors of cultural, social, biological, and physical environment systems. (Clancy 2009: 28)

Cognition is what happens when we sit on the chair.

Some scholars push the boundary on the dynamic nature of selves and cognition in regard to the relationship between perception and action, which are inextricably linked in situated cognition models. Consciousness 'is about acting', it emerges from our behaviour (Prinz 2009: 434). Noë and Anthony Chemero, among others, go so far as to argue that there is no perception without action: 'Perception is not something that happens to us, or in us. It is something we do. ... The world makes itself available to the perceiver through physical movement and interaction' (Noë 2004: 1). This shifts what it means to be in space and time. Action occurs in space and time, and in relationship to others in that space. From this perspective, our sense of space is not first of all about our place, but about our activities within it (Gallagher 2009: 41). It is not 'a *spaciality of position,* but a *spaciality of situation*' (Gallagher 2009: 42, italics in original); that is, space is about doing something.

This linkage of perception and action leads to a provocatively useful troubling of the idea of representation. Representation is a contested term in situated cognition and can refer to mental images, references and neuronal firing patterns, among other things (Gallagher 2009: 47). In previous generations of cognitive science, it was assumed that external elements in our world are represented in our brains. That's certainly what it seems like: thinking of an apple seems to generate a mental representation of an apple. This does not mean, however, that there is a place where all things apple reside in our brain. What apple are you picturing it? From what angle? Is it partially eaten? It is true that the visual system is activated when we think of an apple – which accounts for the fact that we can see an apple in our 'mind's eye' when we think of an apple – but the apple imagined is specific to the context or the prompt, and it relies on the same system of neurons that sees an apple with the actual eye. This is different from a traditional understanding of representation because it is not abstracted and it is not independent from the perceptual system.

Cognitive psychologist Rolf Zwaan has conducted various experiments about the relationship between mental representation and language comprehension. In studies, subjects are quicker to identify a picture of something if it is preceded by a sentence that describes the object in a way that matches the visual stimulation. For example, they showed subjects two pictures of a nail, one horizontal and one vertical. Subjects who first read a sentence that matched the image to come ('she picked up the nail off of the floor' or 'he hammered the nail into the floor') identified the nail more quickly than those shown a non-matching sentence. In other words, in order to understand the sentence, subjects recruit their visual system – imagining a nail lying on the floor versus one upright – and that prime facilitates recognition. Along with the work of others in this area (Barsalou 1999; Gallese and Lakoff 2005; Glenberg and Kaschak 2002 and MacWhinney 2005), Zwaan's work makes explicit the profoundly embodied nature of cognition.

Just because our language separates doing and thinking and feeling and sensing does not mean that's what is going on. Just because it feels sometimes like those are the categories of our processing – I walk down the street, I think about my interview, I feel anxious, I sense the weather change – does not mean those divisions remain useful in making sense of what and where cognition is. This is important, obviously, because it gets to the very core of who we think we are – not just as performers or spectators, but as human beings. We think with and through our moving, perceiving body. Just as we use our visual system to process 'apple', we use our motor cortex for action understanding. The degree to which humans have mirror neurons may still be under investigation, but the fact of our motor resonances is clear. When we witness an actor make an intentional action – grab the gun, cock his fist, lean forwards to kiss her son – neurons fire in us that also fire when we do that action. This does not mean the same neurons, but rather a small subset of the neurons involved in the action also fire in perception. When we read a sentence about an action, it will activate some of the same neurons in the brain required to do the action. For example, Zwaan and Taylor found that, to understand a sentence such as 'he turned down the volume', we will recruit some of the same motor neurons responsible for making the knob rotation action with the hand, even though it is the intention described in the sentence, not the action needed to accomplish the goal. According to Zwaan and Taylor, 'Comprehension does not involve the activation of abstract and amodal mental representations but rather the activation of traces of perceptual and motor experience' (Zwaan and Taylor 2006: 9).

Paradigms of action are increasingly superseding paradigms of (mental) representation, possibly since cognitive processes are hybrid, '[straddling] both internal and external forms of information processing' (Rowlands 2009: 127). It is the primed muscles in the hand, ready to turn down the volume, that are involved in 'representing' our comprehension of the sentence. As Rowlands puts it, 'Representation and action are, indeed, essentially connected – because *acting can be a form of representing.* Representation ... is representation all the way out' (130–31, ital. ours). To have a representation is, in some ways, to act. As Noë pointedly asks, if a person is *in* the world, why does she need to produce 'internal representations good enough to enable [her], so to speak, to act as if the world were not immediately present?' (Noë 2004: 22). We don't need a representation independent of the actions required of us by the object: no ' "translation" or transfer is necessary because it is already accomplished in the embodied perception itself, and is already intersubjective' (Noë 2004: 80). Most radically, Anthony Chemero says that it's all pretty much action and perception, with no need for representation at all; his book, *Radical Embodied Cognitive Science*, is devoted to the argument that cognition is about agent–environment relationships, rather than representation or computation.

As intuitively right as radical views of extended cognition and 'conflated' action and perception might seem, however, it is important to note that some cognitive scientists reject the idea of extended cognition as framed by the dynamical systems perspective, one argument being that

> the standard argument for pushing the boundary of cognition beyond the individual organism rests on conflating the metaphysically important distinction between causation and constitution. ... It is one thing to say that cognitive activity involves systematic causal interactivity with things outside the head, and it is quite another to say that those things instantiate cognitive properties or undergo cognitive processes. (Robbins and Aydede: 9)

This is a prime example of the need for rigorous conceptual and evidentiary clarity in order to use the research meaningfully.

Goals and caveats

This project has immense implications for our understanding of what it means not just to have, but *be* a body; for a start, these fields engage imagination, emotion, constructions of self, interrelationships, memory and consciousness. Current cognitive science dislocates a number of things, among them familiar constructs of identity, feeling and selfhood that have been dominant for decades, and any belief that culture and biology are separable. At the same time, we must be cautious about our applications – even appropriations – of the science and honest about our motives for doing so. Cognitive science is not a monolith. Not everyone agrees, and even those who are persuaded by the arguments understand that good science is always awaiting falsification. As we've pointed out before, however, provisional does not mean untrue; gravity is a provisional theory and yet we don't each feel the need to walk off a building to accept it.

Science has long informed our engagement with theatre and performance, but there are limits to the relationship and caveats about applications of the former to the latter. Scientists use a reductive approach – verification through repeatable experimentation that accounts for all of the variables involved in the experiment; the process is inductive. In the arts and humanities, theorizing (or, more accurately in science's vocabulary, hypothesizing) is typically deductive, based upon the examination of texts and historical artefacts, the observation of performances or the assessment of the experiential through a particular critical, philosophical or political framework, which often involves a good degree of subjective interpretation. The inductive nature of science demands that it be reductive, look at specific, even microscopic objects such as 'a single type of neuron in a specific part of the brain' (Lutterbie 2011: 8). Underpinning the sciences are principles of falsifiability and repeatability. Theories and things presented by science as facts are proven by results that are repeatable, experiment after experiment, and that are always subject to being disproved when a new experiment produces different results; it can be easy to forget that science is highly contingent and involves its own kinds of subjective assessments of evidence. This is of course foreign to performance studies revelling in the complex and contradictory aspects of our objects of study. Interestingly, when scientists bring together the results of a broad range of experiments to reach general conclusions, more variables come into play across the experiments and often make conclusions more speculative, that is, the broader the assertion, the further scientists move into conjecture and hypothesis and away from science. For this reason, among others, those of us in performance and theatre studies must engage primary and secondary scientific research, educating ourselves in the terrain of these sciences and the standards by which they operate. Summary articles in recognized journals are useful points of entry because they provide a context for and condensation of research on particular topics, including competing arguments and claims; we must engage or at the very least acknowledge these competing claims if we are to have a hope of being responsible in our appropriations and applications.

Different kinds of evidence, ranging from the neural to the linguistic and the behavioural, are useful for different aspects of performance and theatre studies but cannot be applied whole cloth. Among the things those of us working with this material have been learning are: we have to be cautious in using research on the neural level to explain anything in the realm of the experiential or conscious; for example, the discovery of mirror neurons in monkeys did not immediately mean that humans had mirror neurons or that they functioned identically in humans, or that the discovery of neural simulation in humans meant that we are intrinsically empathetic. We have to be clear about what is presented as scientific theory, that is, an explanation that accounts for observable phenomena, following processes of repeatability and falsifiability, and what is speculation, that is, *possible* explanations for phenomena that have not yet been borne out by experimentation. It is important to discern between data and conjecture. We have to be sensitive to contradictions and disagreements among the scientists' explanations of what they have discovered by experimentation and how they are interpreting it. We have to be clear about the differences among the various cognitive science disciplines in terms of methodology and parameters for truth claims, for example, there are differences in the processes and perspectives of cognitive linguists and neuroscientists. We have to be respectful of the power – conscious or otherwise – of metaphor and the intrinsic human tendency to think metaphorically and analogically. Sometimes these associational leaps are apt, and sometimes not, growing as they do out of experience, habit and desire. Having made this last caveat and acknowledging that the use of science *qua* science can be profoundly important, the use of science as a springboard to engage theatre, dance and performance can be incredibly rich – but this potential for creativity and for experiential and intellectual efficacy is different than making a claim that what we do has the efficacy or 'truth' of science.

A topography of where we've been

The proof may be, as Bertolt Brecht said, in the pudding. We have found the integration of scientific research into our studies of theatre and performance useful in how we think about our work, and that is most important. Each of us has discovered new ways of working with text, with performance, with actors and with reception through the interplay of arts and science. We believe that it is a programme of inquiry, not a space race. In other words, what is important is working together to hone and improve the work, not being the first one to make a particular claim or use a new scientific discovery. This requires reading and responding to each other's work. It is no longer enough to know the theatre background and the science background, you also have to know the history of those scholars who have made the connections before you. Some may have done it in a way that you find problematic. Some may have failed to appreciate a key factor. We hope the writers in this book and the readers of this book will continue to move the field forwards by building on what's happened, challenging false steps. To that end, we will try to map a broad topography of the terrain of cognitive scientific approaches to the arts and humanities.[2]

If the terrain is cognitive scientific approaches to the arts and humanities, the work is vast and varied. At this point in the history of the integration, there are even different types of cognitive science that scholars turn to. Some cognitive science theories used in the arts and humanities are less embodied than others; Lisa Zunshine's work on literature and theory of mind (2008), Frederick Luis Aldama's approach to narrative theory (2010) and Paul Armstrong's phenomenological examination of the work of literature on the brain (2013), for example, are centred on thinking about the mental experience of reading. Although they probably wouldn't agree that their work is 'disembodied' (nor are they likely to agree that they even belong in the same category as each other), their focus on reading makes it easy for the body to recede. The cognitive poetics works of Peter Stockwell (2012) and Reuven Tsur (2012) have embraced a more embodied approach to thinking about reading and poetry – though they both have different takes on 'cognitive poetics'. Ellen Spolsky has made extraordinary contributions to the field from *Gaps in Nature* (1993) to *Contracts of Fiction* (2015). Mary Thomas Crane's work on Shakespeare (2001) attends to the performance event, but she is also most interested in what cognitive science opens up for her in terms of thinking of a historically situated brain. Here she shares important contributions to history and new historicism with Alan Richardson (2001 and 2010). Barbara Dancygier (2012) writes about stories from a cognitive linguistic perspective, one that presumes an embodied mind, and, as in her chapter in this volume, she thinks about language as embodied and multimodal. For those of us writing about text, it is important to think about our work in the context of (at this point) a fairly long tradition.

Work in theatre and performance dates from a bit later than the earliest work integrating cognitive science with literature. John Emigh turned to neuroscience to help him think about masks and teaching during the 1990s (Emigh 1996 and 2002). Bruce McConachie's influential and playful 'Doing Things with Image Schemas: The Cognitive Turn in Theatre Studies and the Problem of Experience for Historians' (2001) looked at drama in history through the lens of the work on image schemas coming from Lakoff and Johnson and others. Rhonda Blair's work thinking about Stanislavsky-based acting and cognitive science started with an essay in *Method Acting Reconsidered* (2000). Since then the work has expanded. As evidenced in this book, there are many different questions to ask about theatre and performance that research in the sciences might help address. How can we think differently about design, given what we know about our visual system (Di Benedetto 2010)? How can cognitive science shift the questions we ask about theatre and performance in history (McConachie 2003; Nellhaus 2010; Carlson 2010; Stevenson 2010; Tribble 2011)? How can it improve our methods of actor training (Blair 2008; Lutterbie 2011; Kemp 2012)? What are the different kinds of questions we might ask about audience perception and the experience of going to the theatre (McConachie and Hart 2007; McConachie 2008; Cook 2010; Rokotnitz 2011)? How might we reimagine the situation of the performer/environment interaction (Paavolainen 2012)? In what ways might theatre and performance provide new ways of treating people with autism (Shaughnessy 2012 and 2013) or memory loss (Noice 2006) or Parkinson's (Modugno et al. 2010)? If the quality of the papers at many of the conferences and in many journals is any indication, there is much more excellent scholarship to come.

The book in your hands

This book brings together some exciting work being done at the intersection of theatre and cognitive science. The sections – Languages, Bodies, Ecologies – organize the work by area of focus and methodology. These are not discrete zones, of course; you cannot have any one of these things without the others, but the categories narrow the topic and allow for more specific, rigorous interventions. Each section begins with a short introduction that situates the chapters in the context of the cognitive science research from which they come. We aim to make explicit the methods and potential of interdisciplinary praxis and to increase literacy across the disciplines. While the chapters can certainly stand alone, the introductions provide a guide for those interested in that area of focus but new to the methodologies deployed within the chapters. Each section then concludes with a response chapter from an important scientist/philosopher. In this way, we hope to model the kind of conversation that can take place between those studying particular areas within 'embodied cognition' and 'situated cognition', and those of us experimenting with the implications for theatre and performance in the laboratory of our rehearsal and performance spaces.

The three categories (Languages, Bodies, Ecologies) reflect three key points of attention of the theatre and performance scholar. While these attentional shifts require slightly different methodological toolkits, the sciences being referenced across the three sections are in conversation. In other words, although Vera Tobin's analysis of irony in the Languages section does not explicitly reference the body in the theatrical system, the cognitive linguistic research she turns to is committed to an embodied mind. The three topic areas are deliberately broad, allowing authors to engage more specific research areas as they apply within a context focused generally on language, the body or environments and ecologies. Arranged in this order, the topic areas proceed from a focus on language and meaning-making, that is, ways in which we organize our relationship to our environment; to a focus on the body, the material source of these; to a focus on dynamic ecologies of which the body is a part.

Language is where we reach out from ourselves and describe, label and attempt to change the world around us. Research in cognitive linguistics has been tremendously influential to theatre and performance studies in part because the methods of analysis and the sources of evidence are fairly similar. Our dialogues start with this altered and exciting understanding of language.

Our second section moves to bodies in performance, and how work from the laboratory is being used to explain, expand and improve work in the rehearsal room and in performance. This is an area of growing excitement and energy; since theatre practitioners have always spoken of the 'wisdom of the body', the science is adding new information and new energy to old instincts. Just as we cannot separate the brain from the body – it is *part of* the body – we cannot separate the human from his/her environment.

The Ecology section integrates research from philosophy and the sciences that situates the cognitive process in the interactions of a group, not just in and between individuals, but spread across humans in a particular environment.

The 'After words' is a series of 'origin stories' and interviews with practitioners. We wanted to make explicit the paths of those of us who went from being practitioners to being scholars integrating cognitive science with theatre and performance because we remain committed to the improvement and expansion of our discipline, rather than proselytizing for the sciences. Our stories are then echoed in our interview with John Emigh, emeritus professor at Brown and an early adapter of scientific research, and in the essay written by Deb Margolin, playwright and original member of Split Britches. We spoke to professional artists about their work and how they think about the reception process. We hear in these voices the artistic instinct and the urgency of the practical realities of the rehearsal room.

Finally, we include an appendix with brief abstracts of some of the influential scientific and philosophical works referenced in the chapters. Written by the authors who cited them, the abstracts situate the research in relation to the arts and humanities. We hope this offers readers an additional opportunity to familiarize themselves with some of the science shaping our work.

The book in your hands can be read page by page, front to back, or back to front: searched from the Index to find key areas of individual interest. As with anything, how you use it depends on who you are. Professors might use the abstracts and chapters to inspire students to think about evidence and argument. Artists might be interested in starting with the responses from the artists and practitioners at the end and then tracing an interest articulated there to a chapter that might expand upon it. For example, Tristan Sharps talks about his 'site-responsive' theatrical works (he prefers this term to 'site-specific') in terms of an audience member's response to changes in his environment: How might Evelyn Tribble's chapter or Shaun Gallagher's response help us understand the work Sharps is creating and why? Actors might be interested in how a different way of thinking about their body and craft might give them additional tools. For example, how might Edward Warburton's research on dance help rethink how one memorizes lines? How might understanding embodiment help an actor 'get out of his head' and start thinking smarter? Researchers interested in cognitive approaches to the arts and humanities might hone their strategies by thinking about the different ways the chapters included here do interdisciplinary work. Some scholars will come to an essay in this book as part of a research on 'irony', 'multimodality', 'Julius Caesar', 'gestures', 'actor training', 'flow', 'marking in dance', 'ArtsCross', 'Dubstep', 'The Admirable Crichton', 'expertise' or 'MasterChef'. Others may start with the origin stories, interested in how theory, practice and the practicalities of one's life contribute to finding one's way on an interdisciplinary path. We hope that wherever you start, you find your way through.

Part One

Cognitive Linguistics, Theatre and Performance

Language is a necessary prerequisite for theatre and performance. All performances, even those without language, are attempts to communicate, to gesture through the flames as Antonin Artaud might have it; thus, studying theatre and performance is consonant with a study of how we make meaning through language. Whereas language was thought to come from a language area of the brain with an inherited grammar structure that generates language based on rules, poetry and creative language were considered special cases: deviations from literal meaning. This approach can help solve the analogy section of the SAT, but it fails to capture the richness of Shakespeare or Gertrude Stein. It works well for parsing sentences such as 'the dog walked behind the cat', but fails to explain how we can make sense of 'there's no there there', as Gertrude Stein said about Los Angeles.

Cognitive linguists, philosophers and psychologists began shifting the assumptions about language in the 1980s. Cognitive linguistics generally refers to a branch of linguistics that views language as a reflection of our thinking and not just as the product of a module in our brain that gives words to our thinking, and it has shifted our understanding of how we compose and understand language. Rather than being a system of rules that connects outside 'things' in the world with inside 'ideas' in the mind, language and meaning are embodied and creative. Thus, language is one way to investigate how we think.

There are some key precepts of cognitive science that are central to its integration in the arts and humanities, and which are important for language:

- 'Meaning' is an embodied process. We project information about our experience in our bodies onto more abstract concepts in order to understand the more abstract in terms of the concrete and physical. The state of the body is not only an input into language interpretation, it is also an output. When we read, 'she handed me back the letter', we are much quicker to perform a movement moving our hands towards our bodies than away, for example, suggesting that the comprehension of the sentence accessed the motor cortex sufficiently to prime one physical action (movement towards) rather than another (movement away) (Bergen 2012: 79–80). Others have extended this kind of result to show that the hand muscles are primed even by sentences that describe metaphorical exchanges ('You delegate the responsibilities to Anna') (Glenberg et al. 2008). There is a growing body of research suggesting that comprehending language is a full-bodied affair. We cannot rely on readings that disembody language and talk about meaning as a kind of semiotic code. We require a new look at the theatre that moves us.

- We think and speak metaphorically. There is no literal meaning that receives primary attention. Metaphors and image schemas increase our cognitive efficiency: we use an understanding of one thing (seeing what is before us, for example) to make sense of another (knowing something) such that we can 'see' what another person means.

- Thinking and speaking requires compression. Compression here is a term for the unconscious process by which we reduce the scale of something. This is what happens when we 'turn the accomplishment of many years into an hour glass' as we do when we see a play and know that time may elapse between scenes. Compression is what we do when we see the crown onstage as more than a golden circle or when we point to a map of our town and say: 'here we are'. It is both a fundamental aspect of all cognition and a cornerstone of theatre.

Imagine a character comes onstage through a door up centre, his shoulders hunched, his head bowed, he blows on his hands. A woman enters from stage right and says: 'Come in out of the rain! Let's get you warmed up.' The audience knows that the man entering is cold not because of a translation of the 'signs' the actor gives but because the perception of his physical acts involves some of the same neurons that are activated by those physical actions; thus, through perceiving we are (in part) doing, which is how we know. The woman's line projects the container schema onto the stage area – as we do with rooms and houses – such that one can be in a room or out of a room. The theatrical frame enables us to perceive one part of the stage as 'in' and the other part as 'out', even though there may not be four walls or a roof. We are able to use the container schema to organize many things in our lives as containers that only partially (if at all) resemble a prototypical container. We speak of being *in* a deep sleep or going *out* of our skins. Similarly, there is no elevating the character to get him warmed 'up'; as he gets warmer, he will not begin levitating. We understand that 'up' is good, happy, strong, healthy, positive, and 'down' is bad, sad, weak and negative. Otherwise, how could we understand 'she was so high after her performance', or 'the movie brought me down', or 'she is so above you' and many other such sentences? Compression enables us to decrease cognitive load by reducing the complicated and diffused into the focused and essential.

The three chapters in this section rigorously interrogate the intersection of language and performance from a cognitive linguistic perspective, drawing conclusions that are useful for theatre and literature studies. These chapters do not make cross-cultural assessments of language use; indeed, the plays attended to in these chapters were written by white males. There has been important cross-cultural and multilingual work in cognitive linguistics and we do hope to read more in the future on translation theory and non-western plays. Nonetheless, the chapters make important claims about the way language works onstage. They articulate ways of looking at language and performance, ways that are not specific to a writer or language.

Barbara Dancygier extends the research in cognitive linguistics to the use of objects and bodies onstage to tell a story. Language in the theatre is inherently multimodal and Dancygier explains the importance of incorporating the materiality and, what she calls, the dramatic anchors of the play (such as the props on the table in *Arcadia*) to expand our dramatic analyses. Laura Seymour's chapter examines the way kneeling functions in *Julius Caesar* to perform and stage the up/down metaphor, engaging an audience's empathetic reaction. She focuses on kneeling, both historically and cognitively, and applies her analysis to 'readings' and stagings of the play. Vera Tobin notes, 'There is something about the theatre that makes it seem be a fertile setting for ironies in general, and something about irony that seems distinctively theatrical'; her chapter argues for a kind of theatrical analysis of language, one that sees the rich characters and spatial relations necessary to imagine in order to make sense of ironic expressions. Each combines the nuanced approach to art and literature that a humanities perspective mandates with the paradigm-shifting work in cognitive linguistic research.

Multimodality and Theatre: Material Objects, Bodies and Language

Barbara Dancygier

The nature of theatre as an art form relies crucially on two major channels of expression – which I will refer to as 'modalities'. The modality which aligns theatre with other literary forms is the use of language, but the modality which distinguishes it, and also aligns it with language-independent forms like dance, is the material form which grounds the linguistic expression. There are various dimensions of the material aspect – human bodies, objects, the stage itself, the set, etc., and it is important to ask specific questions about the role of each in constituting the genre. Opening these types of questions is also useful in building stronger connections between cognitive approaches to theatre and cognitive approaches to language. The latter are now undergoing a major shift towards dealing with embodied and visual aspects of communication and bringing these facts to bear on questions regarding the emergence of linguistic meaning. Theatre, as an inherently multimodal form of art, can provide additional strength to these endeavours, by highlighting the power of multimodality and language in their intensely creative form.

The multimodality of theatre

Defining theatre as multimodal art has certain analytical consequences. The description aligns theatre with a range of artefacts where multimodality is central to the form and meaning.

Most naturally, theatre becomes a close relative to opera – where the stage and the bodies are used similarly, but language is downplayed in order to give more importance to music. But contemporary art forms – highbrow, lowbrow and middlebrow alike – rely heavily on multimodality as well. Examples are varied, and new forms keep emerging. Artistic forms may combine visual display, music and poetry reading in various ways, whereas open-air events often mix theatre and music. All these forms combine language and image with specific spatial and visual solutions. In what follows I will look at some examples of interactions between space, material objects and language.

Graffiti art and street art are an interesting example of the treatment of space. They often rely on visual display of language while using visual images in interesting combinations with the material objects and surfaces on which they are displayed. In this sense, street art uses what is available in the material world, and reconstrues it creatively. Such reconstruals often rely on the very basic components of cognitive development, called image schemas. In a recent article, Mandler and Cánovas (2014) argue for several levels of spatial foundations of cognition. They show that in the earliest stages of child development children learn to understand concepts like 'move into', 'occlusion' or 'container', so they have a very early sense of spatial dimensions and accessibility of objects. These schemas then quickly develop into more complex structures, combining with other concepts. Now, street art often plays on these very basic structures, redefining spatial configurations in ways sometimes surprisingly similar to what theatre does. For example, a German street artist, Evol[1] has painted an oblong block of concrete at the edge of the pavement to look like a block of flats – in the ugly style of plain concrete architecture. The work not only relies on the materially available piece of the environment, making it into a representation of a building, but also, through playing cleverly on the difference in size between the painted concrete block and the surrounding buildings, it creates a surprising and attractive effect.

Also, through this artwork, the entire environment has changed its nature. In fact, the first image in the online gallery referenced here has a construction worker sitting on the 'building', talking on his phone, which adds to the surprising effect of size, but also makes the viewer think of another, much smaller city inside the real city, and notice the ways in which such realities may coexist. On the theatrical stage, reconstruals like this one are a natural occurrence – buildings are smaller than they would normally be, while rooms may be a lot bigger, natural environment may be signalled by a single tree, and food is often represented by empty dishes, large enough to be seen from a distance and yet somehow in proportion to the actors' bodies. And yet these token material presences (scenery, props) are intended to totally redefine environment – turn a flat surface of wooden boards into a small part of an entire living world – the storyworld. What street artists often do, then, relying on the combination of the visual and the material, is what theatre also does, in genre-specific ways. And we can argue that the spatial cognition underpinnings underlie both forms of artistic expression.

The primary difference is the approach to containment. Street art creates a space that redefines the surrounding space and blends the two into a more complex and multilayered artefact. In effect, the familiar reality of a street becomes an unbounded stage on which various scenarios can be played. But exactly because it is unbounded, it enriches the experience of the reality space.

Various examples of street art (graffiti or publicly funded installations) have their effect precisely in construing the world as multidimensional. When we see metal sculptures of people walking out from under the pavement[2] (so that parts of their bodies are invisible, under the surface of the street, while the upper parts are already on the surface, among the passers-by), we have to construe an underground space wherein they have existed until now and are at present emerging from. The role of a theatrical piece is to do something very similar – suggesting those invisible worlds of which the audience can only see a part, building the entire narrative icebergs by extrapolating from the visible tips. And yet, assuming the existence of those invisible spaces is necessary for a play to make sense. This holds true for the Classic Greek Stage, Noh Performance, Shakespeare's Globe, Ibsen's drawing room, and the contemporary post-dramatic theatre; a little stands in for a lot.

The stage, unlike a city environment, is bounded. However, the boundaries it imposes are of two kinds. The permeable areas of backstage and offstage are like the invisible spaces from which the street art figures are emerging. Their crucial characteristic is that the audience does not see them but has to imagine some sort of continuity of the onstage world. When a character enters, in the context of the play, he or she enters from the space belonging to the space constructed on the stage – another location, another room, etc. Knowing that there is no actual adjacent extension of the onstage space does not change the fact that viewers treat incoming actors/characters neither as aliens from outer space nor as stagehands who have come to remove a no-longer-needed prop. These persons come from another part of the storyworld.

The other boundary, the one between players and audience, is more impermeable, because the audience is not supposed to enter the storyworld. Of course, contemporary theatre does many things to create an impression that there is in fact no boundary between the world of the viewers and the world of the stage, but it is only successful to a degree – the boundary may be a see-through one, and, so to speak, hear-through or even touch-through one, but it does not go away, and the sense of conceptual independence of the real world and the storyworld remains valid. Still, the manipulations of the conceptualization of space rely on some shared mechanisms. The spatial multimodality of a theatrical event relies on some of the same mechanisms that other multimodal forms display as well.

There is also the issue of how we interpret actual discourse on the stage. In ordinary forms of spontaneous communication, linguists talk about a 'deictic centre' – the here-and-now of the discourse, in which the speaker always refers to herself as 'I' and addresses the hearer as 'you'; of course, these roles alternate throughout the conversation.[3] But the set-up regulates much of the understanding of what people say – so, if a speaker talks about something happening 'now', it is naturally understood to relate to the time in which the conversation takes place. Much work has been done on showing that using such expressions may also be a token of the speaker's viewpoint. For example, in the discourse analysed in Rubba (1996) speakers talk about 'here' when they talk about a (pretty distant) neighbourhood they align themselves with, rather than the room or the building in which they are being interviewed. There is thus a natural understanding that the here-and-now does not have to be the actual here-and-now. In theatre, the deictic centre of the audience is naturally different from that of the play, but the viewers can align themselves mentally with the location and time they are observing.

Thus, in theatre, the deictic centre is that of the storyworld, not the actual reality.[4] In other words, *Romeo and Juliet* always takes place in Verona, regardless of whether it is staged in London, Munich, Tokyo or Verona itself. The storyworld is the here-and-now of the play. But the 'I' and 'you' represents the essence of the play's multimodal nature – the bodies and voices of persons from a different world (that of the current out-of-theatre reality) are used to represent the speakers and addressees in the storyworld. But, at the same time, the words spoken in the storyworld are addressed to the viewers in the out-of-theatre reality. Acting is a living blend of two realities – the content of what is said represents a character and the storyworld she or he is part of, but the body and the voice come from a different reality. But its effects are also bi-modal, as the words inform both the storyworld characters and the viewers, often affecting them in different ways. Other material tokens of 'character-hood', such as costume, hairdo, make-up, etc., are also material blends – they belong to the storyworld and the reality, both. That seamless connection between the body and the voice on the one hand and the words spoken on the other represents two realities, but the audience responds to the 'I's and 'you's spoken as representative of the storyworld – because language users cannot but automatically accept the words spoken to be the actual words of the speaker. The deictic blend of theatrical discourse is what creates the illusion of reality on stage.

The multimodality of theatre poses important questions about the potential role of each of the modalities and about the nature of the interaction between them. As the instances of art forms discussed above clearly show, communicative modalities do their job regardless of the context they are used in, but interaction between modalities can reveal or suppress aspects of the construal. There are various ways in which creative work puts multiple modalities to work, but theatre can be understood better if that broader context is taken into consideration. This chapter looks at some of the dimensions of multimodality of theatre and discusses some examples from drama to illustrate the points. Initially, though, I will outline some of the aspects of both the linguistic and material modalities that play a central role in the construction of meanings. The overview will also show the central role of the human body in linking the two major modalities and allowing the audience to grasp the meaning of the events. The body is the focus of the disturbing blend created, and it is also the vortex of the multimodal complexities of theatre as an art form.

Theatre and narrative structure

Drama tells a story. The narrative structure is complex and significantly different from other narrative forms. Traditionally, theatrical events tell stories in a more or less sequential way, and the changes of location are not easy. I will not elaborate here on the traditional requirement of the unity of place and time, but we can speculate that narrative verisimilitude may have been part of the motivation. Taking deixis seriously could lead to the understanding that theatre provides an opportunity for people to witness the events and hear the conversations as they occur, here and now, so the audience is easily imagined in the role of a silent and invisible witness to the actual events unfolding on the stage. The occasional need to step outside of the 'witness' mode and become an addressee created the role of the chorus, which was a theatrical equivalent of the omniscient narrator.

These specificities aside, for very practical reasons it is difficult to tell a story through drama which would be as scattered as a contemporary novel might, unless the playwright resorts to innovative solutions such as Stoppard's *Arcadia*, and has the present and past realities on the stage simultaneously. Another solution is using one of the characters as a narrator on occasion – but then the assumption is that the character is recounting events that he or she saw, so the 'deictic truth' is still the core assumption – the speaker tells the story to the addressee here and now, reporting on 'then' and 'there' when relevant, but at the time of the events, the current teller was an observer just like the audience is now. This, as the example from *Julius Caesar* (discussed below) shows, opens interesting multimodal questions: What if the audience needs to experience realities off the stage, not just learn about them? The offstage – the extension of the storyworld onstage – is inaccessible to the audience's witness role, and thus needs to be narrated, implied or guessed. Because the story is mostly acted out and only occasionally narrated, the use of modalities has to be different from that of a novel, when everything is relayed through language.

Arcadia is a very relevant example of how complex narrative structure is built through the constant onstage location. The play takes place in one room – the schoolroom in Sidley Park, a large country house in Derbyshire. Some of the events belong to the past of the house, 1809, while other scenes take place in our times. In the middle of the room there is a large table which accumulates a number of objects as the play goes on. It starts with some books, a portfolio and a sleepy tortoise. Stoppard's stage directions as to what is present in the room and on the table are very rich in detail, although he also adds: 'During the course of the play the table collects this and that, and where an object in one scene would be an anachronism in another (say a coffee mug) it is simply deemed to have become invisible' (1993: 15). What these directions do not provide is a clear answer to the question – invisible to whom? Clearly, they are to be invisible to characters from the other period, but are fully visible to the audience. The audience needs to see and appreciate all of them.

The objects on the table are the central focus of the events – a book of poetry, gardening books, an architect's portfolio are all left on the table in 1809 scenes, to be then picked up and looked at by the characters 180 years later. The tortoise is always there. Early in the play, at the end of a scene, a character in 1809 leaves an apple on the table, and at the beginning of the next scene a character in the other period picks it up and cuts out a slice to feed it to the tortoise. The piece of fruit thus becomes a material 'connective tissue' linking the past and the present into one story. It is interesting that the first such instance is of an object that is so easy for the audience to notice and recognize. Other objects that transfer between times are ones containing crucial information. The play touches upon a broad range of issues of history of literature, physics and mathematics, but the plot, essentially a mystery plot, depends entirely on how characters in the present discover the story of the past from documents left behind, scattered and lost.

All the objects on the table are material 'carriers' of information – they provide easy access to the thoughts of the characters who wrote their notes and letters, read from books and produced drawings. They are also crucial to the narrative construction, as the mystery plot is untangled entirely on the basis of the various documents, as the contemporary characters gradually uncover their content. Importantly, the viewer gets most of the story, except its dramatic ending, from the 1809 scenes, while the contemporary characters learn gradually, acquiring evidence and making mistakes. In other words, the story-construction process, focused on the past, is in a way divided between the audience and the contemporary characters – an ingenuous narrative structure, built out of the scraps of information left on the table.

The materiality of the schoolroom, the table and its mess of objects is the vortex of the play which jumps back and forth across a gap of 180 years, until, in the final scenes, both periods are represented in the room at the same time – the shared location is further reinforced by a blended time. In an earlier work (2012), I have argued that the story emerges as a blend[5] of various narrative spaces. The final scenes of *Arcadia* seem to be giving this idea its material dimension – the physical location of the stage displays the blend.

Besides, *Arcadia* makes a very clear use of space, perhaps *because of* rather than *in spite of* its focus on one room in Sidley Park. There is the Sidley Park garden outside the French windows, and there is a music room next door. These are not seen, but often heard. In many instances, what is heard from the garden or the music room is suggestive of the time other than the one represented in the schoolroom – the story blend is thus maintained throughout, and in various modalities – sound, materiality and location. Additionally, one of the contemporary characters, Gus in one time and Augustus in the other, joins a party in costume appropriate to the 1809 period – as an embodied representation of the blend of the two periods.

To sum up, *Arcadia* exploits all modalities in order to help the viewer construct the story spanning 180 years and involving various mysteries of passion and science. We cannot say the story is 'told' by the play – it is multimodally represented for the viewer to build.

The language of the stage

The interaction between language and other modalities not only defines theatre as a different artistic event, but also changes the language of the play. There is a common assumption, again built entirely on the standard role of the deictic centre, that, at least roughly, discourse on the stage is the type of discourse which is normally appropriate in conversation. Of course, the 'roughly' part is where the whole issue resides. Even in the case of contemporary drama, it would be easy to see that it is not the case. It is enough to compare a corpus of conversational discourse with any play to see that people commonly speak in a much less coherent way, that the turns in conversations are often shorter, that there are numerous false-starts and sudden changes of tack, etc. It is not my goal to seriously compare theatrical discourse to natural discourse, but we can learn quite a bit about the language of the theatre by looking at earlier forms, as in Shakespearean drama.

Poetic devices aside (people did not speak in iambic pentameter and nobody claims that they did), there are aspects of theatrical discourse in Early Modern drama that need to be viewed in the context of the story being told and the needs of the audience. The audience needs to know what happened and what characters tell each other, but they also need to know what is in the characters' minds, and it is hard to count on the characters always communicating their thoughts clearly. What novels started doing through various means of 'representation of consciousness' was also done on the Early Modern stage, but without stylistic devices that emerged in modernist prose.

The linguistic formula that emerges very naturally out of the deictic set-up of any play is not just speaking, but speaking to an addressee. However, it does not have to mean a conversation, but rather a form of speech which is ostensibly addressed to someone, but with no expectation of a conversation-like exchange. There are essentially three forms it might take. The most obvious one of them is a soliloquy – a longer reflection, and most naturally addressed to the audience, though not necessarily involving eye contact with the viewers. The audience understands 'To be or not to be' in the context of a soliloquy; the story world is semi-permeable and Hamlet is talking to us about the events within.[6] The second form is speaking in the presence of addressees, but in a manner so deeply focused on the speaker himself/herself that the presence of others is made legitimate only through some very basic forms of interaction, often not responding at all to what the primary speaker is saying and not creating a two-way exchange. For example, in *Richard II*, the king engages in long pieces of oratory in the presence of his courtiers, but they are not part of any conversational exchange. In Act III, Scene 3, he finds himself on the walls of Flint Castle, in the presence of Duke of Aumerle and Bishop of Carlisle. When Aumerle tells him that Northumberland is returning after talking to Bolingbroke, Richard digresses into a long discourse starting, 'What must the king do now? must he submit? / The king shall do it: must he be deposed?' (3.3.43-4)[7] He deplores the deposition he fears, he weeps over his downfall, but the discourse does not respond to Aumerle's announcement or prepare us for what Richard will do once Bolingbroke arrives. Only in the last four lines does he address Northumberland, who has arrived meanwhile. The whole fragment has two goals – giving Northumberland time to get up to the castle walls and letting the viewers appreciate Richard's thoughts. Ostensibly, he talks to his courtiers, but they do not respond or interact, until the plot picks up again with Northumberland's arrival.

The third form is pretending to actually have an addressee, and speak to objects, ideas, images, elements of the environment, material objects or to oneself.[8] Each of these dialogic structures impacts the meaning made of the language by the audience. In an earlier scene in *Richard II*, Richard has just landed on the coast of Wales, returning to his kingdom. Moved by this, he engages in another monologue-like discourse:

> I weep for joy
> To stand upon my kingdom once again.
> Dear earth, I do salute thee with my hand,

Though rebels wound thee with their horses' hoofs:
As a long-parted mother with her child
Plays fondly with her tears and smiles in meeting,
So, weeping, smiling, greet I thee, my earth.

(3.2.4-10)

Throughout this fragment, Richard is using the earth he stands on as an addressee, saluting his land, and then asking it to not support the feet of his enemies who are trampling on it. This ostensible address again has the role of letting the viewers understand Richard's emotions – love for his land and his revulsion at the invasion. The formula of speaking to a material presence, not a human interlocutor, is Early Modern drama's equivalent of free indirect discourse – even though it takes on the spurious form of direct discourse pattern.

It is particularly important in the context of materiality, which is my focus here, that the faux-addressee is represented by the very stage – in this scene, representing the land off the coast of Wales. In the instances I discussed in earlier work, such as Juliet's dagger ('O happy dagger! This is thy sheath!') or Romeo's poison vial ('Come, bitter conduct! Come, unsavoury guide!'), there are in fact objects – props – being spoken to, held up for the audience to see. But the very wooden boards of the stage are a different case – the material stage on which Richard stands is also the storyworld material surface on which he stands.

A full discussion of these types of onstage interaction is not within the scope of the present chapter, but the very range of such forms suggests that a close analysis of dialogic structures of discourse on the stage is needed if the sources of meaning in the discourse of theatre are to be fully understood. Language is the primary communicative modality in a theatrical performance, but it is not an independent one. The division into the page and the stage cannot be maintained. Language participates in a multimodal set-up and its form is adjusted to suit the needs of the multimodal event unfolding.

The meanings of materiality

Material modalities of the theatre are numerous, and each has a role to play. I have already mentioned the stage as a representation of the storyworld, and elements of the play's scenography also contribute to the representation of the central aspects of the storyworld. But there are also material objects.

Objects are bearers of all kinds of meaning. In his companion book to an exhibition at the British Museum, Neil McGregor (2013) discusses material objects found on the archaeological sites of the theatres of Shakespeare's time.

A fork, a dagger or a cup, they each tell a story. From the cognitive perspective, the story crucially depends on what one knows about the historical context. The fact that the small fork found would have probably been used by a theatregoer to pick oysters out of their shells represents a kind of knowledge structure that is now referred to as a frame (Fillmore 1985 and 2006). Thus, a frame of 'fork' today would simply mean 'an eating utensil', but the presence of an unusually small fork within the bounds of what used to be Rose theatre evokes the kind of knowledge only some historians and archaeologists might have – that people would snack on oysters in a theatre (the rest is understood – eating oysters with your hands would be a very inefficient practice).

However, objects are not always used according to their typical frames, and they can undergo so-called 'functional reframing'. I have somehow never invested in a real rolling pin, but I have washed a wine bottle long ago and I have been using it to roll-out the pie dough ever since. So, when I ask a family member who will be helping me bake to take out the rolling pin, she or he will take out the empty wine bottle, which has been functionally reframed and so can legitimately be called a rolling pin. This may seem like a very down-to-earth process, but it depends on our ability to use words on the basis of new practice or inside knowledge and not solely on the basis of a dictionary definition, standard shape or typical usage. Of course, if I want a bottle of wine open, I would not call it a rolling pin, since the basic meaning has not changed and not all objects normally called 'wine bottles' will now be called 'rolling pins'. This household anecdote, however, describes the potential of objects to play roles other than their basic functions – and this is what theatre builds on.

When no reframing takes place, objects are most naturally used as props – in such cases they would be used in the same way in which theatregoers would use them in their lives. So, if there is a stool on the stage, it is most likely going to be used by a character who is going to be seated, but if there is a throne, it would require the presence of a king. Occasionally, however, material objects begin to play roles beyond the representation of some real-life routines and are reframed in the context of the discourse. In this vein, Sofer (2003) talks about the stage life of props to show how various types of objects take up meaningful spaces in drama. It is important, however, to show how the very nature of the objects interacts with the emergence of the story. When their material role stops being the central one, they cease to be simple props, and become what I will refer to as *dramatic anchors*.

The term 'anchor' is used here following the work of Edwin Hutchins (1995, 2005), who discusses material anchors to complex blends. One of the examples he talks about is a watch, to show how our very complex construal of time-measurement is built into the idea of a watch, with its face representing the cycle of hours in a day, the hands moving to indicate the position in the cycle, etc. An enormous amount of conceptualizations is effectively suppressed from view and compressed into the pretty simple interaction with the face of the clock. Hutchins talks about 'offloading' layers of conceptual structure onto the anchors, so that we do not have to think of the cycle of the day, the length of hours or minutes and all that complexity every time we need to know what the time is. It is enough to remember how important it is for a small child to learn how to read the clock to appreciate the degree of compression the anchor carries. Of course, the compression has gone much further with digital watches, and watches built into phones, so that now many people have entirely compressed the material form of a clock out of their construal of time-measurement.

The fact remains that there are material aspects of our interaction with reality that rely on various layers of compression of conceptual structure. Another example of such a concept is a queue – an abstract idea of organized sequential access which manifests itself through a number of people standing in line, facing in one direction, and progressing towards the desired access point as the line moves. There is no material line anywhere, and yet the conceptualization makes people behave in a specific, line-like way. One could argue that the stage is another form of a material anchor. It has features parallel to the watch anchor (complex construal of the fictional storyworld as a separate space) and those parallel to line anchor (people orient themselves on the stage and next to it in ways that respect the conceptual set-up – as we have seen, they can even talk to it). Overall, we have to share an understanding of the fact that the stage is conceptually independent of the rest of the world we live in, that only those participating in a construal called a play have the true right to be there, but would not speak in their own words, etc. The imaginary line dividing the stage from the audience, regardless of its history and increasing permeability, is still a material anchor separating the real world from the storyworld. The complexity of the compressions is enormous, but the point is very much in agreement with Hutchins' idea – we are offloading the entire concept of theatre onto such material and spatial conventions and we adjust our behaviour accordingly. It is thanks to these anchors that the material modality of theatre regulates how we read what is being done and said. And importantly, if a play is being staged on an open lawn rather than in a theatre complete with the stage, the curtain, etc., the anchors are figuratively there – we still understand the experience based on the conceptualizations, not just their material representations.

I have built on the idea of anchors before, when I proposed the concept of *narrative anchors* – expressions that evoke elements of the storyworld which help the reader construct the major aspects of the story the text tells (Dancygier

2007, 2008, 2012). For example, in a rather simple case, the idea of 'the green light' across the water is what allows us to construe the meaning of *The Great Gatsby* – the distant and underspecified visual marker that encapsulates the lost love. It is just far enough that it has no concrete shape, but it is rich enough in meaning (representing Daisy's new life) to demand complete attention. Narrative anchors are quite often expressions describing a material object – a photograph, an old letter, a key, etc. These are not simply objects (like props) – they allow the story to achieve its coherence and meaning.

Dramatic anchors have a dual function, and they are both material anchors (though only to the ongoing conceptualization of the play's meaning) and narrative anchors, as they contribute to the understanding of the story. They do materially participate in the performance, but they also guide the viewer in constructing the meaning of the play (consider Chekhov's seagull, as one such example).

Another kind of a dramatic anchor in a play is the body, or rather, the viewer's tacit understanding of the nature of movement, embodied responses to stimuli such as cold, pain or exhaustion, but also responses to emotions – crying or laughing, stooping down as a sign of depression, spring in the step as a symptom of health and energy, shivering with excitement, etc. These do not necessarily apply to a specific body of a specific actor, but the viewer brings in a conceptualization of what a body does, what can be done to it and how it may function as a representation of an event. Thus an actor's body, signalling a whole range of physical and emotional responses, becomes an anchor to the embodied and emotional aspects of the events narrated.[9]

In what follows, I will consider two examples from Shakespeare's plays where material objects function as dramatic anchors, not just props, and where embodiment is evoked for purposes central to the play.

Caesar's mantle

Julius Caesar focuses on the events surrounding the death of Caesar, the plot leading to the killing, and its consequences. But one of its more memorable scenes is the scene where citizens gather to be present at Caesar's funeral. Even though he was not involved in the plot, Mark Antony is given permission to speak to the crowd – this is the famous 'I come to bury Caesar not to praise him' speech, in which Antony ends up praising Caesar and turning the hearts of the crowd from condemnation of Caesar's life to the desire to avenge his death.[10] After making it clear that the deeds of the killers were by no means honourable and that Caesar was not an evil usurper as he was claimed to be, Antony turns to Caesar's body, waiting to be buried, and to Caesar's mantle, which he holds up for all to see: 'If you have tears, prepare to shed them now./You all do know this mantle' (3.2.172-3). Then he reminisces about the first time Caesar wore the mantle, after a victorious battle. The mantle is beginning to function as a dramatic anchor, as it starts to represent the man who owned it – the man who had just been brutally murdered. The cloak evokes not only Caesar, but also the glorious past. The crowd is prompted to recall earlier, happier times of the story, rather than be totally immersed in the current flow of accusations and recriminations. A noble move, and appropriate at a funeral.

But then Antony focuses on the current state of the mantle, showing places where the assassins' daggers hit it. There are several wounds, and then Antony gets to the wound inflicted by Brutus:

> Through this the well-beloved Brutus stabb'd;
> And as he pluck'd his cursed steel away,
> Mark how the blood of Caesar follow'd it,
> As rushing out of doors, to be resolved
> If Brutus so unkindly knock'd, or no;
> For Brutus, as you know, was Caesar's angel:
> Judge, O you gods, how dearly Caesar loved him!
> This was the most unkindest cut of all;
> For when the noble Caesar saw him stab,
> Ingratitude, more strong than traitors' arms,
> Quite vanquish'd him: then burst his mighty heart;
>
> (3.2.178-9)

In these lines, Antony is telling the citizens how Caesar died, but while displaying holes and bloody marks on the cloak he is in fact re-enacting the murder for them. They are expected to no longer treat the garment as a simple cloak, Caesar's possession, but as a body, being tormented and wounded in front of their very eyes, by the hands of men who claimed their deed was honourable. Unlike the past of the previously recalled time of glorious victories, this is a recent memory, so fresh that the blood on the cloak may not have completely dried yet. The immediacy makes it more charged.

By using the mantle as a material metonymic device Antony is displaying the scene of murder, calling out the killers' names and showing the gruesome deeds in detail. Most importantly, though, he describes this scene of suffering and cruelty, but does not claim the horrific wounds to be the ultimate cause of Caesar's death. He suffered from the stab wounds, but died out of horror and disappointment at the betrayal of his most beloved Brutus. His description of Caesar in agony over the betrayal, not physical pain, is the most emotionally charged aspect of his speech.

So how does that happen? Why do the citizens cry for Caesar and want to avenge him, when a moment ago they believed he was a cruel tyrant and deserved to die? Of course, they seem to be unusually susceptible to rhetoric, but besides that they saw a performance worthy of the best stage, and their minds responded as the performer intended. Antony starts with the frame of Caesar – the victorious general. He evokes past events, events which are in stark contrast with the accusations of abuse of power. Not everything the play tells happens on stage, so the narrative is here enriched with this flashback, but also a different light is shed on the personality of Caesar. Using the mantle as a metonymic device, Antony conjures up the body of Caesar, in his military glory. But then the same proud body is hurt in cruel ways, and by showing the gashes on the cloak, Antony is in fact showing the gashes in the flesh. He re-enacts the murder scene, not by aligning himself with the killers (this would require acting out the blows), but by pointing to the bodily harms they caused. In effect, he makes the citizens experience the killing right then and there. They are either taking the viewpoint of horrified witnesses or mentally simulating their own vicarious experience of being brutally hurt, first physically and then emotionally. Mental simulation is often described as a source of gut-level understanding of the events being represented – if that is indeed the case, the viewers' reaction, both in the scene and in the theatre, is to align their own bodily potential with the body evoked by the mantle and imagine being stabbed, hurt and also betrayed. Antony is evoking the body to prompt an embodied reaction; he is using a material object (the mantle–material anchor) to evoke past events and recent events of the story (it becomes a narrative anchor) and to simulate embodied experience to prompt empathy and anger, and to give a different interpretation of the recent event – it was not an elimination of a tyrant, it was a brutal killing and a betrayal by the most trusted friend (now the mantle also provides a focus for the interpretation of events in the play, and so becomes a dramatic anchor).

The scene continues in a rather surprising tone. Antony is aware of the effect he has achieved (which obviously he wanted to achieve) and sees the citizens weeping with pity. He makes a rather unusual comment: 'Kind souls, what, weep you when you but behold / Our Caesar's vesture wounded? Look you here, / Here is himself, marr'd, as you see, with traitors' (3.2.198-200). This sounds almost as if Antony were dismayed with the force of the spectacle he had staged. He is observing that, strangely enough, people are moved to tears by his re-enactment of a killing, but not by the sight of the wounded, dead body of Caesar. He may just be a sophisticated enough orator to know that a powerful re-enactment of murder is more effective than observing the inactive effect of the crime. He (Shakespeare?) may also be enough of an intuitive cognitivist to understand that provoking a simulated alignment with the pain and despair of the great man as he is being killed by those closest to him is more emotionally effective than a display of his mutilated and degraded body. But the fact that the comment is made, almost sarcastically, sounds like an acknowledgement of the multimodal power of theatre. Performance is better than static display, objects immersed in events can speak more convincingly than bodies, perception of simulated action is better than only hearing a story. When he later describes himself as having 'neither wit, nor words, nor worth,/Action, nor utterance, nor the power of speech', (3.2.224-5) we almost want to cry out – but you know how to use the power of a material anchor and you understand the force of performance; you are a man of the theatre.

Richard's crown and the broken mirror

Unlike the swift and brutal death and deposition of Caesar, Richard II requires a long and painful play to be un-kinged (I use the word instead of 'depose', because much more than loss of title is happening to Richard during the play). As I suggested in earlier work (Dancygier 2012, Chapter 6), Richard's deposition is a process that starts early in the play, even when Bolingbroke still feels obliged to kneel in the presence of his king, and is very consistently built into the up/down opposition, in words, the use of the stage, and people's behaviour. Metaphorically, the opposition is typically understood to evoke a whole range of oppositions: good/bad, happy/ depressed, healthy/unhealthy, powerful/powerless or having/not having status. Importantly, the contrast is said to develop into this complex network of opposing meanings on the basis of our embodied sense of well-being. We are active, moving and in control when our body takes the upright posture, and passive, static and unable to control things when we are lying down. One of the most common metaphorical extensions of this essential embodied pattern is into the domain of power and status – the more powerful the person, the higher they are in the hierarchy imposed onto a field of human activity, be it politics, wealth, administrative structure, artistic prestige, etc. It is thus almost natural to see Richard throughout the play talking about himself as moving down. He imagines himself being dead and buried under the surface of the road, as low as one can sink, so that even the lowliest subject can trample on his kingly, then-still-crowned head. He descends from the castle walls to meet Bolingbroke saying, 'Down, down I come', and even though he is actually moving downwards, the meaning obviously signals his premonition of the deposition that will be forced onto him by Bolingbroke.

There are many places in the play where up/down is the central dimension evoked, in all the range of its metaphorical meanings. And in Richard's story the downward spiral presents him as unhappy, powerless, unhealthy, morally suspect and, eventually, dead. The whole spectrum of 'down' meanings is there, built into language, the material form of the stage, movement, use of space, gesture – all the modalities of the performance.

Then comes the deposition scene. Everything is clear, and nobody has any doubts about Richard's fate, but the formal deed needs to be done. Importantly, Richard needs to cooperate in this. Thus the Duke of York tells him he should give up the crown out of his 'own good will', even though he is in fact being brutally removed. Richard is very much aware of the mockery of his own good will in the act and so decides to make it into a performance. In this, he will use the one material object that physically will pass from the current king to the next one, as a symbol of the transfer of the title – the crown. He thus holds the crown and asks Bolingbroke to hold it on the other side, and so they stand, both holding on to the crown (and the title), while Richard talks about his grief. As a performative choice, it is a very strange solution, and not a very dignified one. The two men, two kings, are holding on to a crown, but neither of them is wearing it, nor is there any sort of ritual transfer, though one might expect that Richard would take the crown off his head (where it marks him as king, while in his hand it does not) and an appropriate official would put it on Henry's head. This would not be an official coronation, but a good enough performance (with the use material objects, of course) of the transfer of power.

In the actual scene the two men stand there for a bit, very unnaturally, and so narratively this is a moment of stasis. It is not unusual that an unnatural moment of inaction is built into a play to extend the narrative time taken up by an important event. What naturally can take seconds, may take minutes on the stage, so that the importance of the event is appreciated. Meanwhile, Richard evokes several important embodied dimensions – up and down (again), weight, motion and visibility:

> Now is this golden crown like a deep well
> That owes two buckets, filling one another,
> The emptier ever dancing in the air,
> The other down, unseen and full of water:
> That bucket down and full of tears am I,
> Drinking my griefs, whilst you mount up on high.
>
> (4.1.184-9)

The metaphor of the two buckets is almost unseemly considering the high stakes and the importance of both men. But Richard plays on the oppositions which depict Bolingbroke's success and his own degradation. Richard's bucket is down, but because it is in the well, it is also invisible, and immobilized by its own weight, while Bolingbroke's bucket is up high, light and 'dancing in the air'. Importantly, the image Richard creates is not that of transfer (which would be expected), and the up/down dimension is only the result of the contrast between being free of the burdens of life and being weighed down by them. Additionally, the movement of the two buckets (and men) is connected – when Richard goes down, Bolingbroke goes up. One could argue that even though throughout the play Richard has been portraying himself consistently as 'down', here he is primarily presenting himself as powerless, overburdened by grief, removed from the attention of the world in which he used to exist and connected in a mirror-image way with the fate of the man who wants to destroy him. The crown is the cause of all this, but, for now at least, Richard and Henry are in a political sense in the same situation – they are both holding on to the crown, but they are not kings, either one of them. One is on his way up, the other on his way down. But while they both hold the crown, they share it, and as Richard describes himself to be the king of his griefs, they have to share those too.

The meaning potential of the two objects evoked in the scene – the crown which is materially there and the well with its two buckets imagined by Richard is constructed through evocation of the frames, through exploiting spatial and embodied schemas to build metaphorical meanings and through the inaction and the awkward situation of the main characters. As in other cases, the crucial scene achieves its meaning multimodally – through language, the bodies and their posture, material objects (whether actual or imagined) and the strange lack of motion and action.

In the next part of the scene, Richard calls for a mirror: 'Let it command a mirror hither straight, / That it may show me what a face I have, / Since it is bankrupt of his majesty' (4.1.65-7). The scene that unfolds, at the end of which Richard shatters the mirror, has often been talked about as a closure of the drama of Richard's two bodies – the body natural and the body politic. The shattering of the mirror is then read as the final and irrevocable destruction of Richard's body politic – his existence as a king. I want to suggest that the meaning here is much more complex. Starting with the very meaning of the mirror (described in detail in Cook 2010), we need to consider the material aspects of what the object does and could mean. A mirror gives a reflection of the object placed in front of it – an image that should faithfully represent the reality. This literal meaning is often exploited based on the conceptualization that a representation of reality could be construed as a different type of reality. In literature, mirrors are often used to exploit the idea that when a person looks in the mirror, they see the image of themselves that they normally do not see, for obvious reasons – one cannot see one's own face. So in various novels, characters are described looking in the mirror and seeing someone else – a parent whose influence affects their life, a person they would like to be or the person others would want to see in them. So yes, it is natural to think about mirrors as not mere reflectors of reality.

In the play, Richard wants to see his face. When asked to read deposition documents while the mirror is being brought, he refuses, saying he will read 'himself', like a book, when looking in the mirror. He looks to see if the change in his situation has changed him in a visible way. Metaphorically, however, vision is very often used to represent knowledge and understanding, so we can assume that Richard needs to understand the degree to which the changes in his life affected him as a person. Moreover, the repeated use of word 'face' is important here, in terms of the aspects of embodiment involved. Richard does not just talk about seeing 'himself' – the whole person, body and soul. The face is the central representation of the person, one which reflects not only age and looks, but also any emotional expression. We know people's feelings often from their body posture, but the face is the best indicator of the inner person. What Richard seems to be trying to see in the mirror is precisely that – the inner person that he cannot otherwise observe. In a metaphorical sense, he tries to look himself in the eye.

The material features of the mirror have an important role to play as well, as the final comment Richard makes is about fragility. He talks about the fragility of worldly glory, but also of the face again – the reflection of his inner person he has just looked at: 'A brittle glory shineth in this face: / As brittle as the glory is the face' (4.1.287-8). Having said this, he shatters the mirror, and the reflection of his inner self along with it. The looking glass anchor thus relies on the material features of the mirror – its ability to show what we otherwise cannot see, its ability to reveal something about the person who looks into it, and its fragility. That fragility, or the ability to fall apart easily is what mirrors share with some construals of human psychology. We can talk about a person being 'shattered' to mean that their emotional persona has suffered. The self is often talked about as a brittle object as well (we often say 'I'm falling apart', or 'I'm in pieces'). Based on these features we can interpret the mirror as a means Richard uses to think about himself from his own personal perspective (to 'read' his own face), not from the 'royal' perspective everyone else takes. He also talks about the 'face' (the inner self) being brittle. In its final moment, when the mirror is shattered, the scene gives a material form to the idea of his spirit being shattered and no longer unified, coherent and whole. Under this interpretation, the mirror is a dramatic anchor that completes our understanding of Richard's story and refocuses it from the very obvious political level to an equally important personal level. Richard as a former king stands there, humiliated, while the other Richard, the inner self that has suffered so much, is now broken beyond repair.

To sum up, the material presence and nature of the mirror evokes a range of meanings that make it in to an effective narrative anchor. The onstage performance of its potential transforms it further into a dramatic anchor – an element of the play that builds on its multimodality and material nature to construct narrative and literary meaning.

Conclusion

Multimodality of theatre is the source of not only its meaning potential, but also of its complexity. The interaction between the material, the embodied and the linguistic is intricate, and relies on a number of dimensions: visual perception, frame evocation, conceptualization of the human body, understanding of space and, last but not least, the language. But the use of language in a play is not only an independent though important modality, it also supports other modalities in very specific ways. Highlighting the frames, clarifying the nature of events observed and foregrounding the metaphorical meanings of material occurrences are just some of the roles language plays. In future work, we should attempt to untangle the net of various relationships among the modalities. But also, we may find inspiration in other forms of creativity where multimodality is the primary feature of meaning construction.

Doth Not Brutus Bootless Kneel? Kneeling, Cognition and Destructive Plasticity in Shakespeare's *Julius Caesar*[1]

Laura Seymour

Shakespeare's *Julius Caesar* (1599) involves a significant moment of kneeling: the conspirators kneel to Caesar before killing him.[2] Strikingly, this moment is not in Shakespeare's source, Thomas North's translation of Plutarch's *Life of Julius Caesar* (1595: Xxxiv).[3] Yet Shakespeare takes care to emphasize that his conspirators do indeed all kneel before Caesar. Cassius, for instance, says to Caesar, 'as low as to thy foot doth Cassius fall' (3.1.56). Up until now, the conspirators' kneeling gesture has never been the central focus of studies of *Julius Caesar*. However, theories of embodied cognition suggest that it is right at the heart of the play's concern with power and resistance to hegemony. Because kneeling was so crucial to creating and cementing early modern social hierarchies, it was also a prime location for those hierarchies to be troubled and disrupted. Reading *Julius Caesar* alongside cognitive theories of how bodily movement shapes thought, this chapter explores how Shakespeare exploits this fact, making the conspirators' kneeling gesture the hinge upon which the play turns. This chapter also suggests that paying attention to theories of embodied cognition leads modern audiences to a sometimes troubling new awareness of their own postures as they crane and watch the play.

At first sight it seems counter-intuitive that, precisely at the moment that they aim to destroy Caesar as an authority figure, the conspirators perform a gesture that, early modern writers and cognitive theorists both suggest, has the power to generate submissive, obedient thoughts in the conspirators' minds. Kneeling was a very common gesture of deference in the early modern era. Higher and lower bodies were not merely secondary signs of higher and lower status, rather, kneeling gestures were also thought to be able to create feelings, and relationships, of submission. In their seminal *Philosophy in the Flesh* (1999), Lakoff and Johnson argue that our abstract concepts are inherently based on, and shaped by, embodied metaphors and embodied experience. A bended back, and a kneeling knee then, are fundamental to our idea of what it means to be socially 'lower' than someone. In her Lakoff- and Johnson-inspired analysis of up–down and near–far imagery in *Julius Caesar* (2004), Eve Sweetser (2004: 26–7) shows that with this imagery Shakespeare 'evokes extremely basic shared aspects of our cognitive structure – and yet constantly shows us how deeply ambiguous are the ways they play out in the broader contexts of our culture-specific worlds'. In *The Literary Mind* (1996: 66), Mark Turner examines kneeling in Shakespeare's *King John* as a situated gesture, whereby people who were kneeling were not allowed to forget that they themselves were lower than heaven and the sky. While, in *Shakespearean Neuroplay* (2010: 135), Amy Cook discusses the power of submissive gestures, kneeling in particular, to generate feelings of submissiveness on the part of the gesturer, and to reflect and generate hierarchical power relationships. She writes, 'Who we are – what we feel, what we do, where we are, and what we remember – is then best seen as an embodied, embedded, and transactional *performance*.' By 'performance', Cook says that she means that gestures do not just represent thought but can also produce it, 'the performance of the action does not signify; it creates'. Barbara Dancygier's chapter in this volume ranges from up–down patterns in Shakespeare's *Richard II* to close analysis of Caesar's mantle as an object with a cognitive history, demonstrating both that kneeling is a wide Shakespearean preoccupation and that perspective is just one of many cognitive effects at work in *Julius Caesar*. While Sweetser concentrates on Portia kneeling to Brutus, this chapter turns its attention to the conspirators' moment of kneeling, aiming to establish, as Sweetser does, that cognitive theory re-unearths early modern ways of understanding *Julius Caesar* and also generates new ones.

Julius Caesar is a play that is very attentive to the nuanced, altered social relationships created and represented by characters' different heights relative to each other. For example, Brutus imagines Caesar climbing other people like a 'ladder', then looking down on them from his vantage point, 'scorning the base degrees / By which he did ascend' (2.1.23-8). Cassius describes the conspirators as exaggeratedly smaller beings, no higher than Caesar's legs, playing upon the contemporary dual sense of 'petty' as 'physically small' and 'of low social importance or rank', or (as Cassius goes on to say), 'dishonorable'[4]:

he doth bestride the narrow world
Like a Colossus, and we petty men
Walk under his huge legs, and peep about
To find ourselves dishonorable graves
(1.2.135-8)

As the conspirators kneel to Caesar, both parties draw attention to the social implications of the gesture. We have seen that Cassius addresses Caesar, 'As low as to thy foot doth Cassius fall' (3.1.56). The word 'falling' throughout the play signifies low social status; at Brutus' suggestion that Caesar has epilepsy, 'the falling sickness', Cassius responds with a pun on being content with a dishonourable life, 'No, Caesar hath it not: but you and I ... we have the falling sickness' (1.2.252-4). Before dying, Caesar compares his high 'rank', 'true-fix'd ... quality', and superior social 'place' to 'the northern star', high up in the 'firmament' (3.1.60-9), but when he dies, Caesar registers the end of his authority as his body's physical descent, 'fall, Caesar' (3.1.77). That Caesar's physical fall is precisely the moment of his 'fall' from authority is also registered by Antony, whose first words contrast Caesar's prostrate form with his previous political might, 'O mighty Caesar! Dost thou lie so low?' (3.1.148). Caesar also falls down earlier in the play, though we do not get to see it: Caska reports that Caesar 'fell down' in front of the crowd who are applauding his decision to refuse a crown, and upon reviving 'offer[ed] them his throat to cut' (1.2.250-264). This moment, when Caesar drops to the floor with the effort of refusing a kingly crown, cements the association between falling, vulnerability and low social status. Brutus's new status as high up in the social hierarchy is underscored when he is knelt to straight after killing Caesar; Antony's servant enters and says, 'Thus, Brutus, did my master bid me kneel' (3.1.123).

In sum, there is a lot of pointed discussion about how high or low people are relative to each other in *Julius Caesar*, and Shakespeare works hard to ensure that we associate standing tall with physical strength and high social status, and falling down with physical vulnerability and a descent through the social hierarchy. So this moment when they kneel to Caesar really is a moment where the conspirators construct themselves (physically and socially) as a lowly set of people, only to rise up (both physically and socially) and overtop Caesar. Though there is an undeniable element of conscious planning to Caesar's murder, an attention to, and manipulation of, the vertical axis of the body is crucial to Caesar's downfall.

Early modern readers, steeped in a culture that associated kneeling with performative ceremonies of feudal allegiance, humble prayer and deference to monarchs and nobles, could not have failed to experience the conspirators' kneeling gesture as a richly resonant moment. At the time that *Julius Caesar* was being written, debates among clergy were becoming particularly heated over whether kneeling could produce submissive thoughts in worshippers' and subjects' minds. Drawing on a centuries-long tradition, the official Elizabethan line was that it could – and did.[5] For modern audiences, though, these resonances are not immediately accessible. Cognitive theory, with its emphasis on the gestural embodiment of social and cultural ideas, enables us once more to regain an awareness of the deep and complex significance of kneeling in *Julius Caesar*. As Shaun Gallagher (2014) explains, embodiment has become the dominant mode of cognitive theory, and consists in an emphasis on the 'lived body', an idea of meaning in which 'body position' is central.

The question still remains, just why do the conspirators kneel to someone that they want to kill? I have chosen as the title for this chapter one of Julius Caesar's final lines, in which he questions the meaning of Brutus' kneeling gesture, 'doth not Brutus bootless kneel?' The word 'bootless' could mean 'useless', 'helpless', 'with no remedy', 'unprofitable' or 'without recompense' (the old English word 'boot' referred to profit, help, legal compensation or atonement for a sin).[6] So here Caesar is suggesting that kneeling is worthless, meaningless and an ineffectual persuasive tool. The line 'doth not Brutus bootless kneel?' highlights the potential for the kneeling gesture's traditional connotations of obedience and deference to authority to be ineffectual in a wider sense. Caesar's question evokes contemporary political anxieties about whether kneeling was a pretence or a real submission to authority, whether it was an efficacious gesture in terms of influencing the gesturer and those around them or whether it was merely an empty symbol of obedience. To kneel bootlessly is for that gesture not to have the power to affect either the mind of the kneeler or the mind of the person being knelt to. To kneel bootlessly thus diminishes the gesture's significance as a mind-shaping tool. But, on the contrary, to kneel bootlessly provides the kneeler with power, enabling them to draw on the gesture's powerful set of cultural significances while remaining insulated from its effects on their own mind.

This chapter argues that in the short term, Brutus does kneel bootlessly: he evokes a gesture of submission only to prove that his thoughts are not entirely submissive towards Caesar by killing him. However, in another sense, *Julius Caesar* shows that kneeling has an inescapable power over Brutus' mind, creating and consolidating a relationship between Brutus and Caesar in which Caesar is the victor, the dominant power to which Brutus submits. Brutus is finally forced to admit, 'O Julius Caesar, thou art mighty yet. / Thy spirit walks abroad and turns our swords / In our own proper entrails' (5.3.100-2).

Flexibility, plasticity and habits of the bended back in early modern texts

In *Julius Caesar*, political machinations, tyranny and social revolution are fundamentally produced, and made real and legible, through bodies bending, bowing, and standing tall: countering uprightness with kneeling, and using the kneeling gesture to bring a tyrant to his knees. How can this be so? Cognitive theorists have often turned to the phenomenological tradition, and particularly to the works of Merleau-Ponty, to explore the tight, dynamic relation between gesture and thought. The philosopher Catherine Malabou echoes Merleau-Ponty's own interest in melding neuroscience and philosophy with her theory of plasticity. Plasticity is a process whereby the body's gestures sculpt neural pathways and thought patterns, while the material mind in its turn shapes the body, controlling its gestures and exploiting their established significances to its own ends. I suggest that this is precisely the sort of thing that is at work in the conspirators' kneeling gesture in *Julius Caesar*.

The philosopher Alphonso Lingis uses Merleau-Ponty's theory of the body schema to explore ideas of posture, uprightness and 'standing tall'. Lingis argues that our gestures are given meaning by their relationship to other people, 'Our "body image" is not an image formed in the privacy of our own imagination: its visible, tangible, and audible shape is held in the gaze and touch of others'. Like Merleau-Ponty, Lingis argues that understanding, responding to and mimicking other people occurs 'not with a concept-generating faculty of our mind', but with the motions of the body. He cites as an example our ability to understand the uprightness of a sequoia trunk only by sensing and replicating this uprightness in our own posture:

> When we look at the sequoias, we do not focus on them by circumscribing their outlines; the width of their towering trunks and the shapes of their sparse leaves appear as the surfacing into visibility of an inner channel of upward thrust. We sense its force and measure its rise with the movement of our eyes and the upright axis of our body. We comprehend this uprightness of their life not with a concept-generating faculty of our mind but with the uprighting aspiration of our vertebrate organism they awaken. (Lingis 1996: 64–5)

Lingis's evocative words here are highly suggestive for a reading of *Julius Caesar* in that he argues that that we can both understand and alter our relations to others simply by adjusting the vertical axis of our body. Guillemette Bolens's cognitive discussion of 'kinaesthetic learning', or 'our human capacity to discern and interpret body movements, body postures, gestures, and facial expressions in real situations as well as in our reception of visual art' (2012: 13), additionally provides a way of conceptualizing early modern prayer gestures. Bolens contends that both because cultures have emphasized kinaesthetic learning and because thoughts are physically embodied in the material structure of the brain, understanding the other is a material and embodied process. She argues that a person's thoughts and sensations are anchored and recorded in the body in 'kinaesthetic memory'; we learn and store memories by imagining and replicating other people's movements with our own bodies.

The Anglican (though Puritan-leaning) bishop of Winchester Lancelot Andrewes (1555–1626) discusses kneeling in his chapter 'Of Outward Reverence in Gods Worship' in *The Pattern of Catechistical Doctrine at Large* (published posthumously in 1650). What is at stake for Andrewes is strikingly similar to what is at stake for cognitive and kinaesthetic theorists like Gallagher, Sweetser, Bolens and Lingis: like them, Andrewes wants to analyse and explain how gestures like kneeling can affect thought processes. Andrewes proposes three reasons for worshippers to kneel in prayer.

> 1. That God may be glorified, as well by the body, which is the external worship, as by the soul and spirit, which is for the internal. 2. That our outward gesture may stir up our souls to their duty, as clothes increase the heat of the body, though they receive their heat at first from the body. Lastly, as to stir up our selves, so to stir up others by our example, that they seeing our reverend behaviour, may fall down with us, and be moved to do that which they see us do, and to glorifie God on our behalf. (1650: Ee[r-v])

Andrewes highlights the symbiotic relationship between thought and gesture that is so crucial to cognitive theory. This is one of those places where cognitive theory's emphasis on embodiment pulls the dust-sheets from ideas that were readily accessible to early modern people, but which readers, audiences and critics have lost sight of in the intervening years. Andrewes' assertion that, though the soul gives the body an initial impetus, it is only when the body is involved that a person can experience a meaningful relationship with God through prayer, can be understood using Malabou's idea of 'the mutual fashioning of soul and body'. As she describes in her book *The Future of Hegel*, this mutual fashioning is at the heart of habitual action; it enables gesture to shape as well as be shaped by the mind, and dissolves the barriers between them:

> Through its power of self-repetition, habit creates in man the condition for the reversibility of psychic and physical attributes. The features of the soul, as they acquire a physical means of expression, cease to function as a separate world, or a 'mysterious inner space'. Similarly, the body, as it is made into an instrument, will no longer act as a natural 'immediate externality' and a 'barrier'. (Malabou 2005: 73, 69)[7]

Andrewes' equation of the body as clothing, or a 'habit' (garment) resonates well with Malabou's idea of habit (custom), beginning as a merely external aspect of our being and gradually working its way inwards until it becomes our intractable second nature. We inhabit a role until it inhabits us. We let our habits (external garment-like bodies) become our habits (internalized customs). Like many early modern writers, Andrewes makes the significant assumption that the sight of a kneeling person will encourage others 'to be moved to do that which they see us do'. However, equally characteristic of the Anglican church (which incorporated a wide variety of dissenting individuals, from Church Papists to ordinary laypeople nostalgic for older, Catholic traditions) was the fact that each worshipper potentially knelt with very different thoughts to their neighbour. Malabou's theory of plasticity helps to add nuances to descriptions like Andrewes', and to place a useful emphasis on the fact that early modern subjects could use 'outward gesture' to 'stir up their souls' to many different forms of 'duty'.

So, the strange moment of kneeling in *Julius Caesar* suggests that kneeling is a gesture that authority figures can use to control us, but it is also one that we can take control of ourselves. Cassius begins thinking about this early on in the play, when he notes with simmering sarcasm, 'Cassius is / A wretched creature and must bend his body / If Caesar carelessly but nod on him' (1.2.118-20). Referring to himself in the third person, and setting off his own supposed 'wretchedness' against Caesar's emphatically nonchalant nod, Cassius is clearly viewing his own kneeling gesture with the detached irony of the experimenter: he tests its significances, getting behind its accepted meanings. In Stuart Burge's film version, several decades ago, Cassius (Richard Johnson) underscored this by accompanying these lines with a sarcastic little bow.

A relatively recent discovery, the fact of the brain's plasticity, enables a specifically neural reading of *Julius Caesar* that was not available to early moderns. However, at the same time, Malabou explicitly draws on an Aristotelian and physiological tradition of thinking about the way that bodily habits can sculpt thought which has, she acknowledges, definitive roots in the early modern era. Early modern subjects' bodies were ideally malleable, rather than the 'stiff-necked' politically rebellious people who are lamented over in official clerical texts. Early moderns also spoke specifically of a 'plastic power' by means of which the body and brain changes and develops, but also by means of which, in the words of the natural philosopher Joseph Glanvill (1661: E6^{r-v}), 'the Soul' is 'the Bodies *Maker*, and the builder of its own house'. Malabou's idea of plasticity is useful for understanding how early modern gestures were both ways in which rulers produced obedient subjects and the ways in which subjects used their bodies to resist these very dominant cultural and religious norms. Malabou links plasticity intrinsically to habit, relying on Marc Jeannerod's neuroscientific work into how habitually using certain synapses when we perform certain actions and have certain thoughts strengthens those synapses and increases their responsiveness. Synapses which are rarely used become increasingly less responsive. Thus, our neural pathways reflect our habitual thoughts and actions, making us increasingly adept at performing those thoughts and actions that we most habitually perform.

Malabou's ideas differ crucially from ideas about 'hard wiring' the brain through repeated bodily action. She asserts that plasticity can fundamentally be a conscious, responsive process whereby we choose how to shape our brains for ourselves rather than passively allowing our habits and culture to shape our 'final' personality and thought patterns. Malabou argues that while plasticity can be, and often is, an unconscious process whereby we allow ourselves to be shaped by the dominant cultural norms of our time, we can also control our own plasticity. By being conscious of how our brain responds plastically to our thoughts and actions, we can choose to cultivate certain ideas and practices that enable us to shape our own brains in ways that resist the dominant culture. For instance, a republican conspirator can kneel to a tyrant while thinking anarchic thoughts; each time he kneels he consolidates this thought pattern, making himself more prone to think in a manner different to submissive feudal norms. Thus, Malabou says, in *What Should We Do With Our Brain?*, plasticity can be both active and passive:

> According to its etymology – from the Greek *plassein*, to mold – the word *plasticity* has two basic senses: it means at once the capacity to *receive form* (clay is called 'plastic', for example) and the capacity to *give form* (as in the plastic arts or in plastic surgery). Talking about the plasticity of the brain thus amounts to thinking of the brain as something modifiable, 'formable', and formative at the same time ... plasticity is also the capacity to annihilate the very form it is able to receive or create ... to talk about the plasticity of the brain means to see in it not only the creator and receiver of form but also an agency of disobedience to every constituted form, a refusal to submit to a model. (2008: 5)

For Malabou, plasticity challenges the notion that the soul or the mind is a person's intrinsic personality, and that the body is an unimportant, external representation of the inner self. The theory of plasticity suggests that the gesturing body can adjust a person's thought patterns and even their personality, making new ways of thinking an entrenched second nature. In *The Future of Hegel*, she sees habit as a form of self-possession, rather than being something that allows others to possess us, and links it to the Latin verb *habere*, 'to have'. Malabou writes, like Lingis, from a philosophical perspective.

However, there is a large body of empirical evidence substantiating her claims, as well as bolstering cognitive theories of habit as a whole. Evelyn Tribble's *Cognition in the Globe* (2011), which explains Renaissance rehearsal practices according to principles of cognitive underload and cognitive ecology, is a key example of this kind of work.

In *Julius Caesar*, we can see that what Malabou calls 'the mutual fashioning of soul and body' through kneeling can be used in two different ways, and this highlights a very big ambiguity in the play, with respect to which the kneeling gesture is a real tipping-point. First, the kneeling gesture is a sign of Caesar's power over the conspirators – habitual abasement makes them more and more lowly in mind, it cements the relationship between them and Caesar as one of lower and higher. But they also attempt to take advantage of their own flexibility, and the adaptability of the meaning of kneeling to shape it, and their minds, to their own ends.

Audiences, perspective and being bowed to in the Globe and beyond

We have seen that *Julius Caesar* is a play that is very aware of the ability of gesture to shape thought, and be shaped by it, often at the very same moment and in different ways. This is characteristic of Shakespeare's drama in general: his plays tend to hinge around arresting, powerful gestural moments: Titus Andronicus offering his severed hand to Aaron; Othello kissing Desdemona before killing her; Hamlet and Laertes fencing, darting in and holding back, as they try (and don't try) to kill each other. However, the attention to the vertical axis of the body in *Julius Caesar* can also prompt audiences to be aware of our own bodily axis. The play gives meaning to, and draws meaning from, our own bended backs and stiff craning necks, the leaning of our shoulders as we stretch over balconies to catch the 'Et Tu Brute' moment. As the work of scholars like John Lutterbie has shown, cognition in the theatre is not limited to what happens onstage. Rather, each performance constitutes a much wider 'dynamic system' involving the mutual interaction of actors, audience and props. Shifts in the focus of audiences' attention affect the entirety of the system.[8] In *Julius Caesar*, depending on where we are sitting, we stretch our necks, bend our backs or lean back to see Brutus's kneeling gesture, our focus can shift partly to our own posture, to the tensions in our own body and to our own position relative to other audience members and the actors: Are we higher or lower than them? Where we are sitting, and how we are sitting, in the theatre affects the significance that the conspirators' kneeling gesture has for us. When we sit in the pit, for instance, do we not feel that Caesar is looking down on us, and we, like the kneeling conspirators are forced to look up to him? How does this impact upon our identification with the characters?

In the earliest performances of *Julius Caesar*, the kinaesthetic class differences within the play, whereby characters' high and low social status is embodied by higher and lower bodily postures, will have been replicated and confirmed by the ways in which the audience was vertically arranged. The most expensive seats were situated higher up in the Renaissance playhouse, and enabled the richer (and usually more noble) patrons of the theatre to enjoy a higher position than the poorer, lower class, groundlings standing in the pit below. However, this effect will have been, potentially, infinitely nuanced. For example, though the higher-class audience members in the higher-up seats will have enjoyed the ability to look down on the action onstage like Caesar atop his metaphorical ladder, they will at times have had to bow their heads and bodies to see the players below them, thus forcing them into a traditional posture of submissiveness. This remains the case to an extent in the modern Globe, where it is still much cheaper to be a groundling than to sit up in the gallery, and the pit is populated with cash-strapped students who have paid £5 to be drizzled on. However, the Globe is somewhat of an exception nowadays; in the modern West End in London, the situation is reversed, with the most expensive seats for the richest people (and usually the press, who are being wooed) also being the lowest, in the stalls. And of course we do not kneel or bow to each other nearly so much in present-day England. The fact that the social mix of audience members is not, in the present day, so neatly hierarchized mirror-fashion by their theatre seats is perhaps the reason why modern directors often seek other ways than the vertical axis of the body to portray the kneeling gesture onstage. Directors have swapped the high–low distinction for other, but no less effective, distinctions such as those between darkness and light, long and short durations of time or fixity and fluidity of movement. In so doing, they have often managed, whether as a main aim or a side-effect of their directorial decisions, to erase the significance of the nuanced differences in seat height among audience members.

In the 400th anniversary production at the Globe in 1999, Mark Rylance created a striking contrast by having Caesar's prone corpse lie at the feet of his towering statue, the latter as 'fix'd' and 'unshak'd' as Caesar had mistakenly thought himself to be.[9] This contrast between fixity and movement will, unlike that between high and low, have been experienced in a similar way by all audience members no matter where they were sitting. Reviewers of this production remarked, too, not on the relative heights of Caesar and the conspirators but rather on the time Caesar took to die. Again, this will have been experienced in the same way, wherever you were sitting. Caesar's death was noticeably drawn out, lasting roughly sixty-seven seconds (from Cassius' line 'speak hands, for me' to the final stab) and involving a sixteen-phase fight sequence whereby Caesar tried to fend off each conspirator in turn.[10] Ignoring the height issue completely, Michael Billington wrote in the *Guardian* that it was the sheer difficulty of killing Caesar quickly and cleanly that showed his power over the conspirators and the durability of his authority; he was 'a robust autocrat who was going down fighting'.[11] In Jan Klata's 2013–14 *Hamlet* with Schauspielhaus Bochum, Claudius's struggle to kneel and pray did not focus on him struggling to move from a standing to a kneeling posture at all. Instead, Klata darkened the entire stage except for a tiny spotlit space stage right. As he attempted to pray, Claudius laboriously attempted to move from the dark space to the light space: this dark-light contrast will again have been experienced in the same way by all members of the audience, wherever they were sitting.[12]

One modern production, however, found a way to play with perspective in a way that enabled a nuanced appreciation of the high–low distinction from within the audience. This is Phyllida Lloyd's 2012/13 production of *Julius Caesar* at the Donmar Warehouse, London. This production managed to find a way around the fact that *Julius Caesar*'s attempts to involve the audience in its high–low distinction does not always translate well onto the modern stage because this play was written for a playhouse where the noblest patrons sat higher than the groundlings, something that is generally reversed in present-day theatres. Lloyd's production was set within a women's prison; the cast played a group of prisoners staging a production of *Julius Caesar*. Lloyd argued that the prison setting ('a world of oppression and violence') resonated with the play's themes of hierarchy, 'By setting it in a prison, we are creating a world in which violence is ever possible, freedom is restricted, power and hierarchy are the meat and drink of every person who is incarcerated; where status is important, and where superstition is rife' (Lloyd 2012: 18).[13] This production focused innovatively on the kneeling gesture when the conspirators knelt to then murdered Caesar (Frances Barber). Lloyd deliberately gave her theatre a levelling effect, with all audience members seated on the same level in rows of prison-style plastic seating, so that she could play with perspective and audience experience.

At Caesar's murder, the company carefully drew the audience's attention to the differences in perspective between the kneeling conspirators and Caesar who sat elevated above them on a chair. Barber swapped seats with an audience member in the front row, meaning that as Caesar died he was facing the same way as the audience. As a result, as the conspirators knelt to Barber, the audience shared Barber's perspective relative to them.

Occasionally turning round to address her lines to audience members in a manner that suggested she expected sympathy and support from them, Barber consolidated this relationship and identification between Caesar and the audience. Simultaneously, via a camera trained on Barber's face, an image of Caesar's threatening countenance was shown on television screens positioned high up on either side of the stage. Thus, as well as experiencing Caesar's viewpoint of the conspirators as they knelt and looked up at him, the spectators could see Caesar's face looking down on them, as if they were in the conspirators' position.[14]

Lloyd's production illustrates how crucial perspective and relative vertical height are to significant moments in the play. As Caesar was murdered, Lloyd enabled the audience to experience the radically different viewpoints of the kneeling conspirators and the upright Caesar, thus inviting the audience to ponder the different allegiances with different characters that these two perspectives might provoke. Fittingly, in this production, after Caesar's death, Barber returned onstage to play a tyrannical prison warden who threatens to stop the amateur production in its tracks: a bold suggestion that after his murder Caesar retains authority over the conspirators ('O Julius Caesar, thou art mighty yet'!).

Returning, to finish, back to the material cultures of 1599, the original performance conditions of *Julius Caesar* meant that this play would very likely have been framed by another significant moment of kneeling: as Tiffany Stern has recently shown in a groundbreaking article, the players will probably have knelt to the audience, and at court performances the monarch, at the end of the play. Stern (2010: 27, 25) describes these moments as liminal ones, where the boundary between the fictional world of the play and the 'real' world of the play is uncertain, as they are often 'moments in which the Epilogue becomes cognizant of the audience and "notices" the monarch' when present. Stern argues that final prayers or bows to monarchs at the end of plays that, like *Julius Caesar*, critique the authority of rulers might be contaminated by some of the anarchic elements in the play, even as the final kneeling or bowing gesture attempts to contain those elements by reasserting the players' deferent relationship towards their rulers and patrons. This was especially the case once Shakespeare's company officially became 'The King's Men' with the accession of James I in 1603. 'That means that though, potentially, whatever king or queen may have been questioned or slaughtered within the fiction, the reigning monarch of the time ruled the end of some versions of Shakespearean drama.' For instance,

> the prayer moment in Shakespeare's *2 Henry VI* suggests that on some occasions at least Shakespeare's play ended its exploration of troubled

kingship, its questioning of everything the monarch stood for, with a rousing, monarchical prayer. Ironies encoded in this text can perhaps be traced to that prayer ... its words are already potentially heralded as meaning their opposite. ... Bringing complicated issues of loyalty to God and to the monarch to bear on whatever play had preceded them, they will (depending on the audience's point of view) have bolstered or ironized the play that they accompanied. (Stern 2010: 28–9)

The Lord Chamberlain's Men (Shakespeare's company before they became The King's Men) often performed at court. If the players bowed or knelt to the monarch at the end of some performances of *Julius Caesar*, their action will have resonated with the conspirators' kneeling to Caesar within the play. The preceding action of the play will potentially draw into question the sincerity of the players' act of kneeling deferentially to the monarch at the end of the play. At the same time, the players' final act of deference will retrospectively comment on the conspirators' hypocritical and seditious kneeling gesture, framing it as something aberrant and wrong.

In present-day theatre, it remains very much the rule for actors to bow before their audiences at the end of a production. Does the way that Shakespeare problematizes genuflection in *Julius Caesar* affect the way we as audiences feel when the actors run onstage to bow in front of us, looking in our direction as their bodies bend? Kneeling or bowing at the end of the play, an early modern player or modern actor retains residual traces of his or her character. Half in and half out of the play, part player finishing the day's work and still part fictional character, the players occupy a space where performance and reality are not ontologically separable but shape and reflect each other. This is perhaps the most dangerous moment of *Julius Caesar*, where the dangerous plasticity of kneeling onstage has the danger of jumping the boundary between the play and the world outside, and where it most needs to be contained.

Are we all ready?

In 2014 I watched a production of *Julius Caesar* in the Bussey Building near where I live in Peckham. Here, the actors sat still as a Narnian Hall of Images around the stage when they were not required. Caesar (Matthew Eades) got up to be killed with an air of resignation, of tension before playing a demanding part. The line that most stuck out was that which he delivers to the conspirators right before the kneeling begins: 'Are we all ready?', spoken as a challenge, almost as if he was chivvying up the conspirators to perform the most important gestures of the play. This line is rarely given prominence in performance and here it had the rare effect of signposting, preparing the audience for something momentous as the conspirators knelt.

Once we become aware of the charged moment of kneeling in *Julius Caesar*, it is hard to unsee it. We become aware of the metaphors of height, bending and falling, structuring our empathetic reactions, and we notice how these metaphors' inherent embodiment becomes richest and most complex as the conspirators kneel to Caesar. Concomitantly, we become aware of our own body, and the social networks in which our trip to the theatre re-embeds us. Cognitive theory makes the process of selecting a seat, and the more arduous and momentous process of designing a theatre, stacking people up or spreading them out particularly significant when it comes to *Julius Caesar*.

With its suggestion that the most important moment of *Julius Caesar* is not necessarily when Caesar dies, but in what happens right before he dies, when the conspirators kneel, cognitive theory paves the way for a wealth of new, vertically nuanced productions of this play. So, are we all ready?

Performance, Irony and Viewpoint in Language

Vera Tobin

Irony and drama

At least since the German Romantics, people have been observing that there seems to be a particular kinship between irony and the theatre. The idea that irony criterially involves some special kind of performance has an even longer history. The etymological roots of irony go back to the stock character of the *eirôn* of ancient Greek comedy, and from there to Aristotle's *Nicomachean Ethics*, where he presents the Socrates of Plato's dialogues as a paradigm example of the type. The *history* of irony is thus intertwined with the history of theatre and performance, but in modern discussions there has often also been the intimation that the *nature* of irony and the *nature* of theatre are especially well fitted to one another.

D. C. Muecke, for instance, observed 'a strong link between irony and drama or the theater' (1969: 40) and argued that 'irony achieves its most striking effects in the theater' (45). G. G. Sedgewick, in his book *Of Irony, Especially in Drama*, proposed that 'the very theater itself … is a sort of ironic convention' ([1935] 2003: 37). Some modern critics have even suggested that irony and drama may be in some fundamental way impossible to disentangle. In his *Grammar of Motives*, Kenneth Burke claimed that drama and irony have a shared 'essence' (517) based in a common logic of dialectic and strategic moments of reversal.

Dramatic irony is the most obviously 'dramatic' sort of irony, and a common observation about irony and drama is that the relationship between actor, character and audience that is inherent to the scene of theatrical performance both invites and enhances dramatic ironies. But, as Manfres Pfister (1988: 55–6) has noted, 'It would be wrong to equate dramatic irony with irony in drama since the latter encompasses an extremely broad spectrum of ironic structures.' It would be equally wrong to suggest that critics' claims for a special relationship between irony and the theatre have stopped at dramatic irony. Other sorts of irony often come in for the same analysis, and this general association has become such a commonplace that it appears in what must be nearly its maximally distilled form in the teaching supplement to Edwin Wilson's widely assigned introductory textbook *The Theatre Experience* (now in its thirteenth edition), as follows:

> Irony: Condition that is the reverse of what we have expected; also, a verbal expression whose intended implication is the opposite of its literal sense. Irony is a device particularly suited to theatre and found in virtually all drama.[1]

There is something about the theatre that makes it seem be a fertile setting for ironies in general, and something about irony that seems distinctively theatrical.

We might wonder whether this kinship in fact springs from one source or from many. Different elements of the theatrical scene seem to contribute variously to quite different sorts of ironic effects. On the one hand, there is the impression produced by the fact that characters we see being performed in some sense do not 'know' that they are 'only' being acted, which gives every staged act a certain element of dramatic irony: audiences always know something important about characters' situations that the characters do not. On the other hand, the fact that acting is a kind of sustained *pretence* lends these same acts a potential ironic knowingness of the sort associated with ironic understatement, Socratic irony or sarcasm. Meanwhile, the physical distance between audience and actors, in which viewers often literally sit in judgement from 'on high', invites a sense of ironic detachment.

Is this multivalenced affiliation between the theatre and the diverse phenomena that have been called 'irony' primarily an accident of historical contingency? Or are there some shared cognitive underpinnings that might help to illuminate and confirm this intuitively evident affinity? The cognitive sciences have largely limited their work on irony to the territory occupied by sarcasm and its close relations – but the kinds of phenomena that have historically interested scholars of literature in general and the theatre in particular show us that this limitation has been truly limiting. I suggest that more attention to the literary and theatrical can help approaches to irony in the cognitive sciences be richer and more complete, while research from cognitive linguistics can also help to explain how and why these observations should hold true.

The most common way of thinking about irony in psychology, linguistics and computer science is as a sort of operation on an underlying sentiment or proposition. A great deal of research in this area is focused on the issue of how sarcasm can be 'decoded' to reveal this underlying intended meaning. But irony, broadly considered, is better described as a way of *construing* an expressed proposition or an observed scene. Acts of ironic understanding in general, including verbal, dramatic and situational ironies, all involve a type of dynamic reconstrual in which attention 'zooms out' from one mental space (the ironized) to a higher viewpoint from which the original is reassessed (the ironic). As we'll see, this way of looking at irony can illuminate the cognitive underpinnings of situational and dramatic ironies as well as verbal irony, and the theatrical qualities they all share.

Varieties of irony and the linguistic tradition

The word 'irony' has, over centuries of use, come to name a strikingly diverse but tantalizingly connected array of phenomena, including sarcasm, cosmic ironies and a certain kind of peculiarly sophisticated or detached attitude, as in the ironic enjoyment of camp (cf. Sontag 1964), with many others between. These flavours of irony are so varied that it is not immediately clear that it is correct to treat them as a single class, though their shared name at the very least tempts us to do so. Research in the cognitive sciences has largely elected to take a very narrow construal of irony – but we can and will do better.

One striking shared feature of all these sorts of irony, although not one that is immediately self-evidently relevant to the theatrical connection, is the juxtaposition of contradictory or incompatible ideas. Not all oppositions are ironic, but all ironies involve opposition. *Situational* or *cosmic ironies*, for example, hinge on the contrast of the actual consequences of some action with its intended results and the means by which they were pursued. Canonical situational ironies feature an act meant to produce some circumstance which instead directly prevents it, or else an act meant to prevent some circumstance but which instead, worse still, brings it about – like stepping to the side to avoid wetting your shoes in a puddle, only to fall into a pond – or a stark contrast between the expectation produced by a set of circumstances and its actual consequence, as epigrammatically captured in Coleridge's 'water, water everywhere, nor any drop to drink'.

Dramatic ironies arise from contrasts between a character's limited knowledge of his situation and the reader's or viewer's greater understanding. Irony can also refer to a particular sensibility rooted in contradiction; the German Romantics specialized in one such approach and often cast irony as the defining feature of the artistic mind. *Romantic irony* is, as Anne Mellor puts it, 'both a philosophical conception of the universe and an artistic program' (1980: 187) – the appreciation of the irreconcilable conflicts of various contradictions in life and philosophy, and a willingness to suspend judgement and resist collapsing the indeterminate or paradoxical. And although they are quite different from Romantic irony, phenomena including camp, kitsch or other sorts of ironic – as opposed to 'earnest' or 'sincere' – enjoyment of objects, aesthetics or activities would also fall into this broad category, in which 'irony' serves to characterize not a piece of discourse or an event, but a mode of appreciation.

Work on irony in the cognitive sciences, linguistics and philosophy of language (e.g. Sperber and Wilson 1981 and 1998; Clark and Gerrig 1984; Kreuz and Glucksberg 1989) has in the meantime come to focus almost exclusively on *verbal irony*: those times when a speaker seems in some sense to say the opposite of what she means, or, as John Haiman (1998) puts it, 'conveys the metamessage "I don't mean this"'. The canonical unifying quality of verbal ironies is that they can apparently be 'decoded' if you recognize that the speaker's actual position and the speaker's ostensible position do not match. In this view, the verbal ironist has a 'true' underlying position to be understood, and failing to recognize the irony will result in a serious misinterpretation of the ironist's remarks.

So, for instance, it is a mistake to think that the annoyed commuter who says 'Oh, that's just *great*' when a passing bus drives through a puddle and splashes her with dirty water is, in fact, expressing delight at this turn of events. To understand the irony you have to decode it, uncovering the fact that she is really expressing annoyance. But situational and dramatic ironies don't have the same kind of coded meaning that verbal ironies do. Failing to recognize the ironies of Oedipus is different from missing the ironic part of a sarcastic remark. In a dramatic irony, there's something to observe (or fail to observe), but nothing to *decode*. You're missing something, certainly, if you don't notice the discrepancies that make a scene or situation ironic, but your interpretation isn't back to front as it is in the case of the classic missed verbal irony.

The issue of misidentifying verbal ironies – failures of decoding – is at front and centre of many current empirical research programmes on irony, especially in natural language processing. The aim of many of these computational projects (e.g. Littman and Mey 1991; Utsumi 1996; Tepperman, Traum and Narayanan 2006) is to provide a way around the kind of interpretative errors associated with verbal irony and to provide tools for 'detecting' sarcasm when it occurs. Eager clients for such tools include companies who hope to get a sense of public sentiment about their products and services, government agencies monitoring 'chatter' online, and anyone else interested in pursuing automatic sentiment analysis. The point is to avoid, as one paper puts it, 'misinterpreting sarcastic statements as literal' (Riloff et al. 2013: 1) by identifying circumstances in which, for example, 'words ...

have a strong polarity but are used sarcastically, which means that the opposite polarity was intended'. Linguists working in this area have generally pursued a more broadly descriptive project, but the typical objective there still has been to identify the features that make an ironic utterance ironic, and to explain how language users recognize and decode those ironies.

But of course not all verbal ironies can be decoded in such a straightforward manner. Gibbs (1986) raises the example of a person who exclaims, with ridiculing aversion, 'I love people who signal' after another driver moves into her lane without signalling. Should we conclude that the speaker hates people who use their turn signals properly? No. Does she love people who don't signal? No again; in fact, she may truly adore people who use their turn signals. The sarcasm inheres in the fact that she has chosen this supremely inappropriate moment to say it: it involves a complex of both verbal and situational discrepancies. Another tricky related form is what the humourist Damon Runyon called 'kidding on the square' (1907): a remark framed as a joke but also meant as a real criticism or jab. Kidding on the square is the weapon of the court jester, a way of getting away with telling dangerous truths under cover of facetiousness. This kind of insincerity qualifies as verbal irony under Haiman's definition – it certainly conveys 'the metamessage "I don't mean this"' – but its viability depends critically on being less than entirely straightforward to decode.

These linguistic accounts also don't have much to say about what makes *situations* ironic, only what makes *people* and *utterances* ironic or sarcastic. This constraint seriously limits their utility for literary and theatrical analysis. The linguist Deirdre Wilson (2006: 1725) takes the view that these apparent limitations are only to be expected, given the fundamentally heterogenous nature of 'irony': 'There is no reason to assume that all these phenomena work in the same way, or that we should be trying to develop a single general theory of irony *tout court* … in other words, irony is not a natural kind.'

This is a reasonable position for a linguist to take, but if we hope to account specifically for the features that verbal irony *does* share with other kinds of irony, we will need to look elsewhere. A better theory of the cognitive and linguistic foundations of irony should ideally explain something about whether and in what way verbal irony relates to other kinds of irony, as well as why irony is sometimes very difficult to pin down. Irony is not always stable and it is not always easy to identify. The ability to recognize sarcasm, for example, appears relatively late in development; children typically can't identify sarcastic remarks reliably at all until about age eight, and it takes a few years more before most can do it reliably in the absence of the strong prosodic cues of a conventionally sarcastic 'tone' (Capelli, Nakagawa and Madden 1990). Even adults very often disagree about the ironic status of a given situation or remark, and they think their intentions are much more transparent than they really are (Keysar 1994 and 2000).

In addition to the juxtaposition of contrasting ideas, one element that seems potentially to unite the various sorts of irony is the presence of some kind of complex viewpoint on a single situation. This is the quality that H. W. Fowler (1926) described as the 'double audience' that distinguishes irony from other sorts of incongruity. The viewpoint account (Tobin and Israel 2012) proposes that this intuitive connection reflects a genuine, shared underlying conceptual structure. In other words, we can connect the sentence-level and discourse-level semantics of verbal ironies with situational and dramatic ironies by thinking about irony as a viewpoint phenomenon. By bringing cognitive approaches together with a fuller and richer idea of what irony might consist of, and when it happens, we can not only gain insight into the riddle that opened this chapter – why is there something about the theatre that seems to be particularly amenable to ironies of all kinds? – but also get a better understanding of some of the trickier kinds of sarcasm that have tripped up earlier linguistic and computational approaches.

Irony as a viewpoint phenomenon

Linguists have appreciated for a long time that many kinds of language are inherently 'viewpointed'.[2] For example, all languages include expressions like *tomorrow, later, upstairs, here, this, sir, you* and *come in*, which incorporate a particular vantage in space, time or social position (among other possibilities) into their meaning. And more recently, work on mental simulation in neuroscience (Barsalou 1999; Bergen and Chang 2005; Barsalou 2010; Bergen 2012) has expanded our understanding of perspective taking in language. Viewpoint turns out to be relevant not only to words that manifestly refer to the spatial, temporal or evaluative perspective of individual language users, but potentially to every aspect of meaning construction in language. Even an apparently viewpoint-neutral sentence like *Marie kicked the football* prompts us to generate inherently viewpointed motor and perceptual simulations of the described events. This means that language users are continually engaged in taking up perspectives other than those of their own personal and immediate experience.

Viewpoint in language may of course also be deployed in service of much more complex representations. We can speak and think of other places and times; we can produce counterfactual conditionals; we can represent the speech and thoughts of others (including imagined others); we can embed perspectives within other perspectives, layer them or blend them in a host of fleeting or extended modes of discourse presentation. We do this sort of thing naturally and continually, but as the configurations of embedded perspectives we try to keep in mind get more complicated, keeping track of their relationships to one another can become very computationally intensive. This interplay between our aptitude for perspective taking with language and its relative complexity can explain quite a lot about irony, how it works and when it tends to happen.

Here is where cognitive linguistics can offer some useful tools to help unify our understanding. The Mental Spaces framework, first proposed by Gilles Fauconnier (1985) and further developed in Fauconnier and Turner's (2000 and 2002) theory of conceptual blending, provides a productive way to represent these viewpoint configurations. Mental spaces are a model of the small, local representations people construct in their minds as they think and talk. In this framework, language is not a true or false representation of the world, but fundamentally a *prompt* for cognitive experience. Any expression is likely to be compatible with many different mental space configurations. The meanings we construct in response to or in the course of producing a given bit of discourse are not only structured by what is said but also may be resolved in part by general pragmatic considerations such as relevance, and in part by our personal predilections, store of background knowledge and other idiosyncratic elements that may be particular to our personality and state of mind in the moment. And because a given mental space is always connected to some cognizer, mental spaces necessarily include viewpoints.

Situational, dramatic and verbal ironies all involve a particular kind of interpretative experience or attitude that comes from a doubled viewpoint, a sense that one has 'stepped back' or zoomed out from one viewpoint to another, more sophisticated view, from which one can gaze, smugly or sympathetically, down upon the original. Wayne Booth describes the experience in this way:

> The process is in some respects more like a leap or climb to a higher
> level than like scratching a surface or plunging deeper. The movement
> is always toward an obscured point that is intended as wiser, wittier,
> more compassionate, subtler, truer, more moral, or at least less
> obviously vulnerable to further irony. (Booth 1974: 36)

This experience of irony can arise when an expressed proposition conflicts with the content of a focused space in a way that leads the conceptualizer to adjust her entire mental space configuration. In order to be counted as ironic, an expressed proposition in some focused space must conflict with the content of an implicit or presupposed proposition in a higher viewpoint, as illustrated in Figure 1.

Event
Viewpoint
Focus
Base/Ground
(BD)

After re-evaluation
and decompression:

Viewpoint (V')
Base
(Day not
beautiful)   ironizing

Ironizing space
contains BD   attention zooms out from putative
                                   base to re-evaluated base

Event   (BD)   ironized
Focus

FALSE with respect to base
IRONIC with respect to base

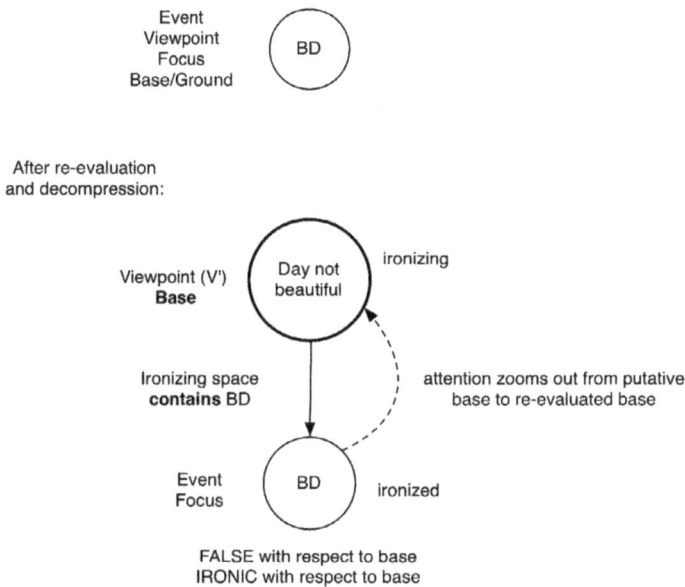

Figure 1 'What a beautiful day'

What you're looking at here is the relationship between two different mental space *networks* or *configurations* representing different construals of the remark 'What a beautiful day'. A little technical terminology: every mental space network canonically includes a Base, a Focus, an Event and a Viewpoint. Sometimes more than one of these roles may be filled by a single space at the same time; sometimes not (Cutrer 1994; Fauconnier 1985). In any case, 'Base' refers to the space that is serving as the subjectively construed ground or basis of interpretation: home base. The 'Focus' space is the space on which attention is currently concentrated. The 'Event' space is the one in which a given event is taking place, and the 'Viewpoint' space is the space from which other spaces are assessed.

Irony depends on the availability or construction of a new Viewpoint space from which one can re-access a formerly in-focus space and its associated viewpoint at the same time. This zooming out both provides the experience of ironic 'distance' and, often by tapping into features of the existing discourse situation, can produce a sense of complicity between the interpreter and some real or imaginary interlocutor. It also captures the sense that many ironies produce of distinguishing between what John Haiman (1998: 80) calls 'the difference between a behaving and a scrutinizing self'.

Figure 1 illustrates a simple zoom-out scenario for a classic example of verbal irony. 'What a beautiful day,' says one person to another, as dismal rain pours down. In achieving an ironic understanding of this statement, the hearer constructs two spaces: a Focus space with the proposition that the weather is beautiful, and a new ironic Viewpoint, which is distinguished from an ordinary observation that the weather is, in fact, not beautiful, by being set up as a higher-level view of the pretended or represented view that the weather is praiseworthy. This viewpoint space represents a new ground and a new potential common ground for communication between the interpreter and some real or imaginary interlocutor.

One nice thing about this approach is that it gives us a way of talking about verbal ironies that connects up with existing accounts of a wide variety of non-ironic elements of language, including verb tense and aspect (Cutrer 1994), conditional constructions (Dancygier and Sweetser 2005), conjunctions (Langacker 2008), co-reference and anaphora (van Hoek 1997) and more. At the same time, it also extends to varieties of irony that go well beyond the examples that are most frequently considered in the cognitive science and linguistics literature.

For example, the zoomed-out viewpoint approach can help to explain the unsettling and unstable nature of many ironies, as well as the related fact that people can and do sometimes reject a putative ironist's own characterization of the ironic status of her remarks, as described by the cultural commentator Lindy West in 2012:

> There's been a lot of talk these last couple of weeks about 'hipster racism' or 'ironic racism' – or, as I like to call it, racism. It's, you know, introducing your black friend as 'my black friend' – as a joke!!! – to show everybody how totally not preoccupied you are with your black friend's blackness. ... Sure, you can't say racist things anymore, but you can *pretend to say them*! Which, it turns out, is pretty much the exact same thing.

In this case, an irony is not missed but vetoed. The experience of irony involves the rejection of the content of a focused mental space and a reconstrual of an original viewpoint space as the new focus. But participants in a discourse may or may not endorse that reconstrual. In the instance of the misfiring ironies described by West, unsympathetic hearers hold the ironist responsible for at least some of the positions associated with the remarks the ironist hoped would be taken as not her own. This doesn't mean, however, that the intended irony simply disappears. An important part of the interpretative experience described here arises from the sense that the 'ironic racist' has attempted to align herself with the hearer in a high-level ironizing view, while the hearer declines to accord the would-be ironist the position she would claim for herself. Something related happens in the case of kidding on the square, in which the ironist actually intends for her putative ironic stance to be rejected by her audience – and yet the resulting construal doesn't (and isn't intended to) collapse to mere sincerity.

What's going on here is that the appreciation of any verbal irony prompts the hearer to decompress the usually invisible blend of expressed viewpoint and speaker viewpoint. In some cases, this may involve first entertaining and then rejecting the literal or sincere interpretation, but the ironic construal can also arise more or less instantaneously. Facial expressions and tone of voice can provide immediate cues for ironic interpretations. People can also approach the act of interpretation with an ironic attitude right from the start, deploying an ironic mental space configuration as a default mode of understanding, as, for example, in the peculiarly sophisticated attitude which takes pleasure in the enjoyment of camp or in the simultaneous appreciation of several mutually exclusive explanations of the world that Schlegel described as the ironic sensibility of Romanticism. In any case, whether an ironic construal is built *à la minute* or pre-compiled as part of an ironic sensibility, the experience consists in the apprehension of two incompatible viewpoints, one of which is rejected and in effect looked down upon. Attention flows from lower to higher. The higher-level common ground from which one views the ironized viewpoint can also already be latent in the discourse situation or genre in which the irony is presented. Novels, for example, often come with a narrator, who is distinct from (though potentially very closely aligned with) the author. Theatre does even more.

Situational ironies and beyond

Irony can arise anywhere that our understanding of a discourse situation provides a multilevel network of mental spaces. Figure 2 illustrates how the viewpoint account extends to cosmic irony.

To die of thirst surrounded by water is to be the victim of a cosmic irony. It is a state of affairs that is tragic, but also absurd, perhaps even faintly ridiculous. To appreciate the irony in such a circumstance requires a certain amount of detachment, a measure of decompression. To take one's own circumstances to be ironic, one must momentarily step outside oneself. Something has happened that might, or could, or should lead to a particular expectation about what would happen. The viewpoint associated with that

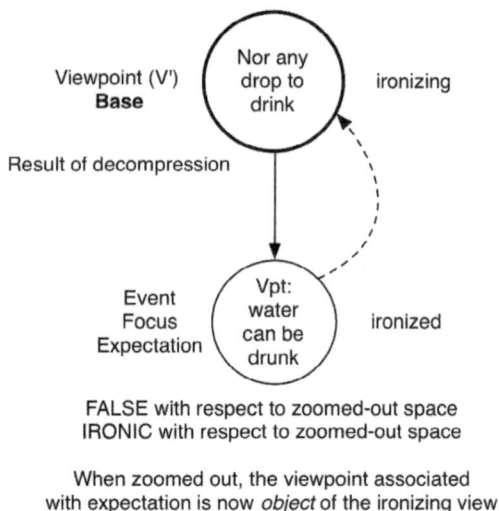

Figure 2 A cosmic irony

expectation – the self who might foolishly think that the last thing to worry about with all this water should be dying of thirst – is the ironized viewpoint, seen from a distance by the ironizing viewpoint.

Dramatic irony, meanwhile, involves a mismatch between facts at the event level and beliefs at a higher narrative level, as illustrated in Figure 3. This example is from Sophocles' *Electra*, as translated and analysed in Sedgewick's *Of Irony: Especially in Drama* ([1935] 2003: 40–2). Here, Clytemnestra's son Orestes has returned in secret from his exile, and sends his attendant, a tutor, to the palace to announce that he is dead. Queen Clytemnestra – but not the members of the audience, who even if they do not already know the story, have seen Orestes set this plan in motion – is deceived. She makes no attempt to hide her delight as she responds, concluding, 'I am freed this day from fear ... now, I say, for all / Her menaces, my days shall pass in peace.' The horrified Electra begs Clytemnestra to curb her exultations, but the queen will have none of it. Sedgewick describes the irony that unfolds in this scene:

> Two opposing courses of action have converged under the spectator's eyes. Clytemnestra's will, purpose, line of action – whatever you like to call it – has long been in conflict (and still is) with the will of her vengeful son. But that any conflict exists any longer, let alone that it means her life – of this the queen is mainly ignorant. Indeed, she exults in a sense of security that she has not felt since she murdered

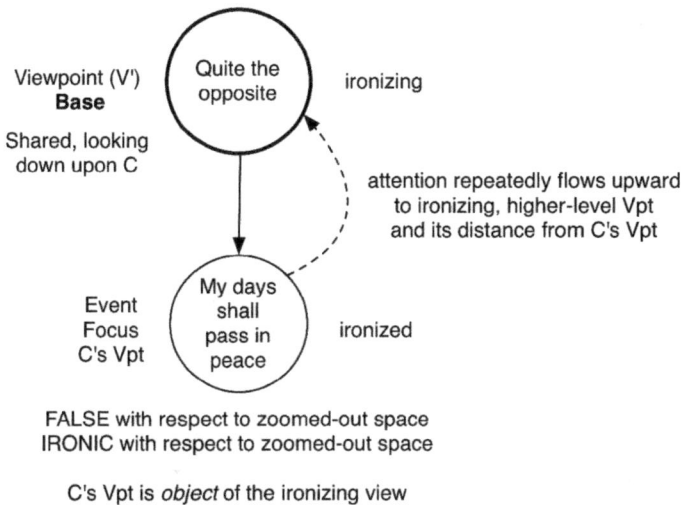

Figure 3 'My days shall pass in peace'

Agamemnon. Her slayer is at hand, and she is welcoming his spy. This is the really *dramatic* ambiguity which the Greek audience certainly perceived and which the Greek dramatists delighted to embody in double-edged speech. (41)

Here the issue is not one of splitting a single perspective into an experiencing and a perceiving self, but of attention flow from a lower- to a higher-level viewpoint. With nearly every line in this episode, the viewer is made increasingly aware of the great distance between the perspective of the character Clytemnestra, on the one hand, and the higher viewpoint shared by almost everyone else (Electra is still in the dark, but will soon be relieved of her ignorance), on the other: our fellow audience members, Orestes, Pylades, the tutor, the playwright who wrote these words, the actors who perform them … all share here together the ironizing view of her ironized viewpoint. The fact that the tutor remains on stage through this scene, listening in and looking on exultantly, sharpens the ironic *frisson*. The ironizing viewpoint is literally in view, and this is not incidental: it in fact (as Sedgewick and others have noted) serves actively to heighten and compound the ironic effect. The tutor embodies this ironizing viewpoint. He gives the audience a perspective that is continually visible on stage, from which the scene is ironic. As viewers, we don't have to hold that perspective in mind, because it is right in front of us.

All of this suggests that irony is fundamentally a matter of how we move around within a particular complex viewpoint configuration. Certain discourse genres provide these sorts of complex viewpoints as a matter of course. Narrative prose, for example, typically establishes several viewpointed layers to which assertions and evaluations may be ascribed – those of characters, narrators, implied authors – and so an irony-ready arrangement is in some sense always ready and waiting to be exploited. In the theatrical context, the viewpoint complexities are physically embodied all around us.

One important element of blending theory is the idea that integration networks very often serve to *compress* information and relationships in the blend: from many elements to few, from diffuse to compact, from between-space relations to within-space relations. The viewpoint account suggests that irony is fundamentally a figure of decompression. More specifically, it is a figure of *desubjectification* – a process in which conceptual contents that are first construed implicitly, as part of the conceptualizer's own perspective, are unpacked and reconstrued as an object of conceptualization 'onstage', seen from the outside. The possibility for irony thus arises naturally from the theatrical mind. However, it also makes high demands on mental processing and, as a result, it typically requires highly ritualized discourse contexts (Haiman 1998).

Thinking is sometimes complex and difficult, and working memory – our ability to keep multiple things in mind at once – is always limited. Because of these limits, demanding tasks that require complex manipulations of conceptual structure generally pose what Edwin Hutchins (2005: 1557) has called 'the problem of conceptual stability'. If we can offload some cognitive labour or inherit it ready-made, allowing us to keep some aspects of a complex network of ideas stable while we attend to others, we can do more complex work with the same working memory. One way to increase conceptual stability is through shared cultural models: the rituals, conventions, institutions and related conceptual and behavioural frames that our culture provides to organize experience. We can also gain conceptual stability by associating conceptual structures with material structures. Physical and cultural structures 'anchor' our thinking and provide scaffolding for complex cognitive work. In the *Electra* example, the tutor is an anchor for the ironic viewpoint on the scene. Furthermore, as Barbara Dancygier describes elsewhere in this volume, the stage itself is a material anchor for the conceptual distinction between real world and storyworld. The physical separation of stage and auditorium (or equivalent) anchors the conceptual separation between the characters and events of the play and the audience who watches them at a safe distance. Indeed, the entire material apparatus of the canonical theatrical setting is ready-made to anchor the experience of ironic interpretation, by physically instantiating the view-of-a-viewpoint configuration that irony invokes.

No wonder irony and the theatre should seem ideally fitted to one another, then. But there's more – once we understand irony as a viewpoint phenomenon and the elements of the theatrical scene as crucial anchors for an ironic viewing stance, the relevance of *acting* to irony also becomes clear. Theatre offers highly salient, always perceptible examples of compressed (but readily decompressible) different viewpoints physically embodied in actors performing roles on stage.[3] As Muecke (1969: 41) observed, while we may know that Hamlet is in some objective sense no more or less 'unaware' of being in Shakespeare's play than 'the daughters of Leucippus are unaware of being in Reubens' painting of their abduction', the actual effect for viewers is quite different: 'Because the Hamlets we see are embodied in actual men whom we see moving and hear talking, it is very easy to think of them as being unaware of their status as actors.' But why should this be, and what does this effect have to do with irony? We can now see that it happens because the usually invisible compression of expressed viewpoint and experienced viewpoint – the same compression that is pulled apart in irony – is continually and visibly manifest in the scene.

The stage, the audience, the actors in their roles, conventions of dramatic structure: all of these help to supply, ready-to-mind, the zoom-out conceptual configuration common to all irony. The theatre makes an apt setting for ironies not just because it is historically or metaphorically linked to irony, but because it supplies a culturally entrenched, embodied context that directly supports the particular kind of complex and cognitively demanding perspective taking that irony invokes. It provides both a material and a cultural anchor for ironic decompression.

A Response

The Performing Mind

Mark Turner

There have been many worthy traditions of analysing multimodal behaviour performed by naturally motivated human beings in real, valid situations. These traditions of analysis are found in rhetoric, philology, philosophy, art criticism and communities of practice. But even in everyday life, analysing performance is routine: when we see the infant, the child, the dinner conversation, adolescent slang, academic status display, flirtation, argument, even the queue of people ordering espresso at the café, we attend to *performance,* and we often analyse that performance. Such behaviours depend on multimodality, on staging, on the use of props for cognition and communication, and on the opportunistic recruitment of aspects of the context to serve temporarily as material anchors. These behaviours depend on gaze, viewpoint, gesture, *performance.* They are what we are built for and how we learn. Any method of inquiry into thought and language not focused on such performances might have been suspect from the start, on several sufficient principles: ecological invalidity, unwarranted assumptions about the nature of causation in complex human behaviour, suspect statistics, the power of unknown intentionality, and so on.

And yet it has taken a long time for the cognitive scientific study of thought and communication to come around to focusing on performance. Even at the impressive pace we have enjoyed since the early 1970s – an iconic moment is Ronald Langacker's radically innovative design for what at the time he called 'Space Grammar' – we still have far to go to recover from the assumptions in which I was trained in the late 1960s and very early 1970s, a formation that included artificial intelligence, logic, mathematics and statistics, syntactic trees, Anglo-American philosophy, human neurobiology and the design of psychological experiments. These assumptions included the idea that all thought was computation of the sort that can be captured on any Turing machine and modelled as such, given enough labour; that the material platform on which the computation was modelled (dolphin, human, silicon or, in John Searle's phrase, beer cans strung up on strings and blowing in the wind) was immaterial; that language was its own independent system and accordingly to be analysed as independent; that causation of human performance was linear rather than complex (in the technical sense) and complied with rudimentary manipulatory theories of causation, where causes are represented as manipulable variables; that we can regress our data to find the linear contribution of a variable to the outcome; that if we understood just a little more about the nuts-and-bolts of our neural machine, it would coalesce into an explanation of mind; that if we could get a computer system to produce outcomes corresponding to such-and-such human behaviour, then the computer program was, implicitly, the answer to the question of how the human did it; and so on. It was a heady and enthusiastic time. I was absorbed by the Zeitgeist for more than a few years. And yet I do not have any excuse: I should have known that all these assumptions were based on the hope of answering age-old questions by ignoring the actual performances that give rise to the questions in the first place.

Jean-Paul Sartre, in *L'Être et le Néant,* famously describes a waiter waiting tables in a French café. Sartre makes the point that the waiter is engaging in everyday life by inhabiting and performing his conception of the role of a waiter, keyed to a particular performance space, which he understands as containing other performers. A little reflection – just the kind of reflection at which I failed in the late 1960s – might lead to the view that anything we do in human life is connected to theatrical performance, complexly, and that the gradient distinctions between official theatre and actual life are sometimes vanishingly small. This can be achieved by thinking of explicitly theatrical performance not as a separate behaviour but rather as occurring in situations designed to bring out aspects of everyday human behaviour, aspects fundamental but mostly invisible to us because we see so little into our own minds and actions.

Consider, for example, the standard kitchen conversation between Person A and Person B. Person A asks Person B what happened at work. Soon, B is raising her voice at A, glaring at A, looking aside and rolling her eyes, pointing a finger between A's eyes, looking up into the air and throwing up her hands, delivering insults, cursing. But A is smiling and nodding, laughing, throwing up hands in imitation while looking down and shaking a head back and forth, etc. From a distance, it might sound a little stressed, certainly look bizarre or prompt concern about domestic disturbance. But, no, A understands, without even realizing that it takes any effort, which parts of the behaviour belong to the remembered boss, which parts to the partner's memory of her performance, which parts to exaggeration, which parts to mock-soliloquy asides commenting on the ongoing report, which parts to the collusive hint to bond as partners or make a martini, and on and on and on. For the readers of this book, I do not need to argue for the connections of this everyday behaviour to theatre. That is probably also the case for any other human behaviour at which we might point. Theatre exists because we already do all of it all the time. From a cognitive scientific perspective, theatre qua theatre is a slightly specialized design for the deployment of pre-existing and ubiquitous theatricality. Inspecting theatrical performance might help us understand human beings better, because the artists have done the hard work needed to haul onstage a little bit of what we find difficult to recognize in the workings of our own performing minds.

Of course, there are distinctions, not to be elided. Actors rehearse and recite lines written by others, while we don't – except that we often do: we often rehearse mentally, even in front of the mirror, or during mental simulations or in daydreams or actual dreams, and in doing so, we borrow lines and constructions and gestures familiar from usage, even if we are unaware of the borrowing. We blend performances we have seen in order to come up with a template for our own action. For various important behaviours, we rehearse, we simulate, we prepare. We might say that actors do not make it up on the fly. The response is that, first, neither do non-actors – human communication is highly conservative and based on usage – and second, yes they do: although one prototype of theatrical performance might be the perfect recitation of something Shakespearean, still, there is a great deal of innovation in such performances (see David Tennant as Richard II, for example), and many performances do depend on improvisation, with the script as only a guideline. We use such guideline scripts often in everyday life. In the loose script for a wedding, the best man proposes a toast. We do not want to lose sight of the distinctions between everyday jabbering and theatrical performance, but neither do we want to mistake them for partitions.

It is perhaps because of the aforementioned shift of focus in the sciences that the arts and humanities are now turning to the cognitive sciences to further investigations into performance. All three of these chapters rigorously interrogate the intersection of language and performance from a cognitive linguistic perspective, drawing conclusions that are useful for theatre and literature studies. Barbara Dancygier's chapter takes the multimodality of theatre as a starting point, explaining the many sources of meaning that come together – from props to environment – when scenes make sense. Laura Seymour's chapter examines the meaning behind the kneeling gesture in *Julius Caesar*, both historically and cognitively. Vera Tobin's chapter on irony and theatre suggests a necessary interdisciplinarity because of the inherent theatricality of irony. When language requires conceptual or mental staging, it's hard to separate it from performance. The ancients often thought of theatrical performance as a laboratory for studying the performing mind. As this book makes clear, we might, too. What scientific methods are available to us?

Methods

The operations of the human mind present the great open scientific problem, but we know that at every moment we butt up against a second, mortifying fact: all methods of investigating the human mind are profoundly indirect and unreliable. No mind has direct access to any other mind, through any method whatsoever. A mind scarcely has any access to itself. The mind seems to be built mostly to not look into itself. Cognitive scientists try to use mental operations that have other foci in order to look at their own operation. What can we do about the fact that all these methods are indirect and unreliable?

In the science of the human mind, we run into a major hurdle confronted nowhere else in science: It is in the nature of the case that people are highly flexible and creative in what they come up with mentally, and they have enormous brains with lots going on and there is no method for knowing in any thorough way what those people are thinking – even they themselves have little idea what they are thinking, and we researchers have no way to read minds. Accordingly, there is no way to predict in the wild what people are going to do or say next. Worse still, there are no good animal models for the higher-order human behaviours we want to investigate, so we cannot perform tests on the animal models and then draw conclusions for people.

The way to avoid pseudoscience is, I think, to embrace straightforwardly the fact that any one method of investigating the mind is irredeemably indirect and incomplete. We should strive to use them all in unison, to supplement and correct each other. Each has strengths and weaknesses. Each calls for repair and confirmation from others. We become more comfortable as more of these indirect methods produce inferences that point in the same direction. Let's consider some of them, all of which I and my research teams have used extensively:

Introspection. Looking into our own mind and gauging our own reactions is an indispensable source of hypotheses, but the many biases of introspection are well known. We tend to think of certain kinds of data instead of others. Our confidence in our introspective judgements has been routinely undermined by analysis. For example, as Dąbrowska (2010) shows, judgements made by linguists diverge from those made by the general population. She concludes that 'syntacticians should not rely on their own intuitions when testing their theories'. The problems of introspection are so well recognized that the subject needs only a cursory mention here.

Behavioural experiment. It is extraordinarily difficult to find ecologically valid contrasts of treatment and control conditions for complex human behaviour. Showing a difference of outcomes between treatment and control in the laboratory might be thought to show something about what happens under the laboratory condition, but leaves unanswered whether the result generalizes beyond the laboratory condition. This is the constant problem of needing to show *external validity*. For example, laboratory conditions led to the conclusion that recognition of false belief by others (the Sally-Anne test) does not take place in children until age four and is not stable until age nine (Wimmer and Perner 1983), but work since then (e.g. Baillargeon, Scott and He 2010) suggests that this conclusion is an artefact of the ecologically invalid condition of asking young children what one doll, or a sketch of a doll, or a sketch of a person, would think about another represented mind. It seems instead that ecologically valid data indicate that infants entering the second year of life already comprehend false beliefs. Similarly, the assertion that understanding of irony is not stable until nine years of age has recently been assessed by researchers who complain that the studies were conducted in laboratory settings, and that when the question was investigated at home, the children seemed to understand and use irony at age four (Recchia et al. 2010). More generally, there is a great deal of activity in any brain at any time, activity we cannot interpret. It may look as if the only difference between a control and treatment group is that in the treatment condition we did one thing but in the control condition we did another. But different people can have great differences in mental activity, so perhaps there are 10,001 differences between them in that moment, not just one. And even if each subject in the experiment experiences both the control and treatment conditions, in a within-subject design, with balancing to mitigate order effects and learning and so on, still, that single subject can have very different mental activity at the two different moments that look so similar, so perhaps again there are 10,001 differences between the control and treatment conditions, not just one. Behavioural experiments are also bedevilled by experimenter effects: subjects, evolved to be highly social intentional agents constantly assessing human preference and expectation, know they are in an experiment, which, it turns out, is often a treatment that swamps all others. The children undergoing the Sally-Anne test, for example, knew some adult was asking them a question and expected an answer. Human beings in social, communicative situations – such as behavioural experiments – are not like chemicals in a tube or electrons in a collider: Human beings are geared to think about the social, communicative situation, and this silent and usually unconscious thinking influences their behaviour pervasively. Other difficulties with assessing behavioural experiments include assuming that human behaviour is the result of simple linear manipulatory causation that can be represented as variables to be measured, even to be located through regression analysis (Woodward 2013). It is also often the case that the subjects have no ecologically valid motivation.

Brain imaging. Brain imaging, an area of great potential if methods improve, remains crude relative to the kinds of phenomena discussed in this section. Event-Related Potential (ERP) has been used to investigate the extent to which a conceptual blend evokes surprise (Nieuwland and Van Berkum 2006), which might tell us something about irony. However, since blending seems to be happening all the time in all conceptual domains, it is difficult at present to see how contrastive methods like functional magnetic resonance imaging (fMRI) could shed light on these processes.

Survey. LaPiere showed conclusively in 1934 that information from survey, polling, questionnaire and self-report can be utterly unrelated to, even diametrically opposed to, what people actually do in ecologically valid situations. Except where a survey or questionnaire has been continually redesigned to correlate with actual behaviour – and moreover it has been shown that this correlation holds, (e.g. for select exit polls immediately after voting) – it seems we can draw no correlations between responses to hypothetical questions and actual human behaviour.

Big ecologically valid data

There is another method we can add to this group of methods to help us move ahead. It is not a replacement of other methods, but a supplement, helping us overcome some of the shortcomings of the other methods. This additional method is the amassing of big ecologically valid data in a database that can be searched and analysed using powerful computational and statistical methods.

To this end, I have collaborated with Francis Steen at UCLA to create a massive database of audiovisual news broadcasts in many languages (English, German, Norwegian, Swedish, Danish, French, Spanish, with others in development). It is called the Distributed Little Red Hen Lab (Google 'Red Hen Lab' to find it). As of the end of 2015, Red Hen had more than 3 billion words and 270,000 hours of recordings, and ingested another 150 hours of recordings per day. Red Hen is able to extract closed-captions and time-stamp them, so they are searchable. Red Hen similarly processes transcripts and on-screen text boxes, and then tags the text for grammar, conceptual frames (using FrameNet) and so on. Red Hen's data are not face-to-face, and not usually in unmediated situations, but they are highly ecologically valid, watched all over the world in nearly all languages by everyone at every age. Red Hen's footage includes not only anchors performing, but also interviews, group conversations, surveillance recordings, crowd-sourced street footage of people who do not know they are being recording, and so on. Hits can be exported to a comma-separated-value file for importing to the statistical software package used by Red Hen, namely R.

So what can a Red Hen researcher contribute to the study of the perform-ing mind? Consider irony. There is a construction in English, often called 'post-sentential Not', in which a sentence is followed by the utterance 'not'. The 'not' indicates that one should embrace the opposite of the meaning prompted for by the preceding sentence. Is it multimodal? Absolutely. Searching for 'post-sentential not' in Red Hen will show how human performers use snorts, head gestures, gaze and prosody with the construction. One can review the performance as often as one likes. Red Hen allows users to tag performances manually, and to search a database for such tags. In the long run, Red Hen would like to use machine learning to do some of the tagging of audio events and gestures automatically. With Red Hen, researchers can study the prosody, the gestures, the pauses and the overall performance in a huge database of performed events to test hypotheses about how these elements impact comprehension. For example, a Red Hen search brings up in a few seconds thousands of examples of kneeling and discussion and analysis of the significance of kneeling.

Red Hen is proud but humble. She is only one database among many. Nonetheless, she is part of a new wave in cognitive science of studying actual performance, and performances of performances (acting). For such a new wave to succeed, we will need researchers in many different fields to work outside of disciplinary silos, with each other, in the kind of cooperation the present edited volume encourages.

Part Two

Bodies in Performance

Bodies, like language, are a prerequisite for theatre and performance. So far as we know, there is no consciousness, no thought, no desire, no action that does not arise from a body. Dancers and actors constantly negotiate these various aspects of the self – mind, body, feeling – to achieve discovery, mastery, confidence, repeatability and, in fortunate moments, inspiration. This section is centred in and on performers' bodies, the material out of which language and meaning-making arise, and through which the thing we call 'consciousness' engages with and relates to others and our environments.

All three of the chapters here draw on cognitive sciences focused on enactivism and embodied cognition; these are particularly apt for performers. These related approaches view mind as embodied – involving the entire organism, not just the brain – and hold that mind even extends into and is constructed by one's milieu: cognition is the result of a dynamic system involving the organism in active relationship to and even in the construction of its environment. Varela, Thompson and Rosch, in *The Embodied Mind: Cognitive Science and Human Experience* (1991), were among the first to use the terms 'enactive' and 'embodied cognition'. These are related to situated cognition, a term that arose in the 1980s to describe how cognition is inseparable from context, or situation, in all of its aspects; informed by psychology, anthropology and sociology, it too is of use in unpacking the complexities of the performance 'ecology'.

The term 'enactive' is a constant reminder that cognition is relational and, well, active. It asserts 'the growing conviction that cognition is not the representation of a pregiven world by a pregiven mind but is rather the enactment of a world and a mind on the basis of a history of the variety of actions that a being in the world performs' (1991: 9). Our experience of the world is *enacted*. It is not based on mental representation or 'objective' measurements. The enactive approach 'consists in two points: (1) perception consists in perceptually guided *action* and (2) cognitive structures emerge from the recurrent sensorimotor patterns that enable action to be perceptually guided' (1991: 173, emphasis ours).

Embodied cognition moves us past binaristic, 'inner-outer', 'objective-subjective' views of experience that rely overmuch on a view of representation as a discrete phase in our behaviour and response. Like enactivism, it resists the compartmentalization of thought, feeling and action and moves us into a holistic vision and understanding of cognition and response; this is crucial for performers. Varela et al. highlight two points in their use of 'embodied' in relation to cognition and action:

> first, that cognition depends upon the kinds of experience that come from having a body with various sensorimotor capacities, and second, that these individual sensorimotor capacities are themselves embedded in a more encompassing biological, psychological, and culture context. By using the term *embodied action* we mean to emphasize once again that sensory and motor processes, perception and action, are fundamentally inseparable in lived cognition. Indeed, the two are not merely contingently linked in individuals; they have also evolved together. (1991: 172–3)

Embodied cognition insists that what we call 'thinking' does not happen in some separate brain but is a function of a living organism in its environment. Within the broader category of embodied cognition there are variations on ideas around simulation, representation, concepts, emotions, etc., but there is general agreement about the two points above and also, as Gibbs articulates in his 'embodiment premise': 'We must not assume cognition to be purely internal, symbolic, computational, and disembodied, but seek out the gross and detailed ways that language and thought are inextricably shaped by embodied action' (2005: 9).

All three of the chapters here take up the inseparability of perception from action, providing examples of the perception–action dyad as a dynamic, ongoing, embodied process of cognition. In short, to use a citation from Evan Thompson that I have used as a touchstone for years, mind is an 'embodied dynamic system in the world', not just a 'neural network in the head' (2007: 10–11). The relationship between us and the world is one of 'dynamic co-emergence' (2007: 60). We are dynamic processes that interact with myriad other dynamic processes.

The chapters here are models for how to apply the science to particular problems in training, rehearsal and performance; they use case studies and examples from studio and performance practice. In each instance, the three authors use the science to illuminate what has sometimes been intuited about how actors and dancers work, strengthening and revising how we might understand and approach the problems of performance.

Neal Utterback, in 'The Olympic Actor: Improving Actor Training and Performance through Sports Psychology,' describes how three techniques from sports – power posing, mental imagery and self-talk – can be used to increase confidence and thereby improve consistency, power and immediacy in rehearsal and performance. His chapter grounds the application of sports technique in the science of embodied cognition. This approach provides concrete tools for actors, directors and teachers on how to use the material of body, vision and language to restructure a sense of self that lets go of negative, 'psychologically' oriented habits of thought, and that serves, rather than inhibits, the actor. Edward Warburton's chapter, 'Becoming elsewhere: *ArtsCross* and the (re)location of performer cognition', is the result of his participation in a multi-year, multisite dance project that involved choreographers, dancers and academics in Beijing, London and Taipei. He presupposes 'dance as a cognitive activity', and uses an enactivist cognitive approach to provide grounding for what has been to this point a generally hypothetical and abstract analysis of how context and situation work in dance. He focuses on memorization and marking, and argues that 'marking' in dance is not just about memorizing and reinscribing movement for the dancer, but it is also about meaning and feeling. His enactive approach to cognition in dance examines cognition 'for action', not just movement, and illuminates dance as a situated activity. Christopher Jackman, in 'Training, insight, and intuition in creative flow', uses an enactivist approach to consider Csizentmihalyi's concept of flow as applied to understanding ease for the actor, describing in terms of cognitive science how rigorous and effortful training can lead to automaticity and flow. He argues that an actor achieves mastery when novel opportunities for engagement are continually afforded through a sustained, dynamic process of problem-solving and finding, that is, when she becomes what Jackman calls 'the spontaneous strategist'. His enactivist perspective reframes skill as 'the very medium by which we come to know our world … reality is disclosed in terms of our own abilities to act upon it'. Taken as a whole, these chapters provide both a foundational range of readings in embodied cognition and a practical, valuable 'toolbox' for how to engage work in the studio and on stage.

4

The Olympic Actor: Improving Actor Training and Performance Through Sports Psychology

Neal Utterback

Introduction

Actors and athletes have a great deal in common, although the two fields rarely cross paths as much as they could or, perhaps, should. In fact, Bertolt Brecht argued that the theatre was lacking in athleticism. In a 1926 letter entitled 'Mehr Guten Sport', the German director and theorist wrote, 'There seems to be nothing to stop the theatre having its own form of "sport"' (Brecht 1957: 6). I would argue that we can take his statement quite literally. Both actors and athletes spend more time and energy in training than they do in competing or performing. Both must contend with a wide variety of stressors, including performing under observation and scrutiny. When striving for professional status, young actors and athletes must learn to self-regulate their emotional and psychological lives and develop personal strategies for staying on target with training and goals. Some sports, such as gymnastics and diving – those that have a great deal of subjectivity in how they are assessed – have even more in common with theatrical performance. Achieving elite status as either an actor or an athlete requires that the individual acquire self-mastery.

What can ultimately separate a novice athlete from an elite competitor of Olympic calibre is the ability to remain cool, focused and confident under tremendous amounts of pressure (Kirschenbaum and Wittrock 1984: 81). The same can be said for the difference between the novice and experienced actor. Any number of factors could cause an otherwise high-level performer to 'choke' under pressure. As Sian Beilock has written, 'Being good at what you do necessitates being able to perform well when it counts the most' (Beilock 2010: 262). That is often when the pressure is most intense.

This chapter is the beginning of a much larger project, 'The Olympic Actor'. Developed from a holistic interplay between exercise science, sports psychology and cognitive science, the Olympic Actor training and conditioning model is geared towards preparing young actor-athletes to become functionally fit, as well as emotionally and psychologically resilient, allowing them to become confident performers and professionals, regardless of the acting technique they ultimately choose. In this initial work, I will detail the first element in the Olympic Actor training model. This three-part intervention involves using power posing combined with mental imagery and positive self-talk. The combination of these three elements into a single pre-theatrical exercise has unique benefits. The goal of this approach – and of the Olympic Actor method as a whole – is to develop a personal scaffold of confidence and positive mood, focus, attention, motivation and arousal.

Let me begin with a case study. I teach at a small liberal arts college in a rural town in central Pennsylvania. Within the theatre department, we have two permanent faculty members. Professional teaching artists supplement our curriculum. Part of my charge is to take new works to festivals around the world, exposing our student actors to professional experiences outside of academe. In 2013, a play that I wrote and directed was listed to be featured at the Dublin International Gay Theatre Festival. Founded by Brian Merriman, the festival is the largest of its kind, bringing together major LGBTQ artists from around the world. In this production, the entire cast of undergraduate students was composed of good-looking, charming, hard-working individuals who were dedicated to becoming better actors. One actor I'll refer to as 'Actor X' had never been out of the state, let alone the country, nor had she participated in a professional theatre festival prior to this opportunity. After one rehearsal, she expressed feelings of anxiety and apprehension. She was worried that she was not up to the task of performing in such a high-profile environment. The anxiety was causing her to perform below her ability level during rehearsal and training. In other words, she was choking under pressure.

The situation of Actor X may be familiar to those who teach, direct and coach young performers. What I began to notice – not only in Actor X, but also in most of my young student performers – was that they often lacked deep, resilient self-confidence and cognitive strategies to mitigate external and internal stressors. Because many of them, like a large part of the student demographic, also lacked proper nutrition and exercise instruction, they were largely sedentary and ate poorly, partially due to the tremendous lack of options in their dining facilities. Moreover, regardless of whether I was using Stanislavski, Meisner or Viewpoints acting techniques, these did little to attend to the actor as an individual. Yes, most approaches assume that an actor should have a strong, supple body and a clear, confident mind. Constantin Stanislavski, in his section on ethics and discipline, outlines the daily demands placed on actors prior to the daily work in the laboratory or rehearsal hall (Stanislavski 2010: 553–80). But even the great master does not build a bridge between the *what* and the *how*. To attend to the rigorous physical demands on the body and the will of the actor, I turned to sports psychology and began consciously to approach the problem of confidence and choking from the perspective of what I call the 'actor-athlete'.

As an acting teacher and director, my goal is to create conditions in which my young actor-athletes are more likely to perform at their best. Like athletes competing for spots on an Olympic team, actor-athletes may be classified into four possible career categories. Olympic trainers Natalie Stambulova, Alexander Stambulova and Urban Johnson see their athletes as winners, failures, reserves and hopefuls. 'Winners' are those who have garnered success in one game and now have the luxury of deciding whether they will pursue a subsequent Olympics. Perceived competence, high levels of self-confidence and the ability to adapt characterize these elite athletes. 'Olympic Games' might be seen as analogical to a professional acting gig, a high-profile audition, or any other activity goal that an actor must engage in frequently, and the parallels become clear. Olympic 'failures' are those who fail to perform optimally or are otherwise outperformed by other athletes. The third group, the 'reserves', are athletes who have yet to compete in the Olympic Games. They are experienced elite athletes who often already have a high national or international ranking, but who could not meet the Olympic requirements, usually because of injury or health issues. They are the understudies on standby to replace the first-string athletes should they be unable to compete. Finally, most teachers and academic directors deal with actors who fall into the fourth group, the hopefuls. Olympic hopefuls have not competed in the Olympic Games but have some experience and competency in their sport. Likewise, their theatrical counterparts have likely had some success in amateur theatre and, possibly, even some professional exposure. However, both groups are still in a significant growth and training phase. Young actors share many qualities with aspiring Olympic athletes who are, according to these Olympic trainers, 'young, ambitious, risk-taking, optimistic, and often immature as personalities' (Stambulova, Stambulova and Johnson 2012: 681). Actor trainers, like Olympic trainers, strive to move those in the 'hopeful' to the 'winners' category.

While Olympic coaches recognize that they need to condition their athletes both inside and outside of the gym, our actor training and conditioning often stops once the class or lab has ended. We spend a great deal of time on the technique and task-specific goals of the actor, but precious little time on the actor's personal psychological resilience and other pre-theatrical cognitive scaffolds. Without psychological and emotional resilience, actors and athletes are more likely to perform sub-optimally (Fletcher and Sarkar 2012: 672). I realized that I needed to train and attend to Actor X – to the pre-rehearsal, pre-laboratory and pre-performance individual – to insure that she did not choke under pressure during rehearsal and performance. Choking is not simply poor performance. It is sub-optimal performance or performing below a standard one usually achieves or is expected to achieve, sometimes due to doubt or over-thinking a task. Beilock writes, 'Choking occurs when people think too much about activities that are usually automatic. This is called "paralysis by analysis". By contrast, people also choke when they are not devoting enough attention to what they are doing and rely on simple or incorrect routines' (5). Beilock outlines a number of strategies for coping with heightened stress. The first approach reaffirms the self-worth of the individual. It is not unusual for young actors to devalue their sense of self-worth or have low self-confidence. They are constantly being asked to examine their bodies and those of their peers; they frequently engage with performance theories that seem mystical or unachievable; and their teachers are also often their directors, so they can be very outer-directed. Added to this is the general stress of being in college. Thus, if we are to engage in the training of young actor-athletes more fully, developing their confidence must be an important component.

Sports psychology can provide helpful tools for increasing confidence. David Fletcher and Mustafa Sarkar note that, for elite athletes such as Olympic competitors, an important difference between success and choking is confidence, which they describe as 'the degree of certainty one possesses about their ability to be successful in sport' (674). When comparing the brains of elite sports athletes and novices, sports scientist Bradley Hatfield found that the elite athletes' brains present less activation. In other words, elite athletes are 'thinking' less than more novice players, who may be consciously doubting their choices and worrying about their skills. As Beilock states, part of what seems to make elite athletes extraordinary is their ability to 'just do it' and bypass conscious thought and control (197). Research in sports psychology supports the argument that confidence accounts for a great deal of success and failure for athletes. In *Sport Psychology: from Theory to Practice*, Mark Anshel writes that being able to maintain confidence levels, both in the state form (i.e. for a particular task) and in the trait form (i.e. generally persistent globally over time), may account for sustained levels of positive emotion, improved concentration and focus, effort and faster and more accurate decision-making (Anshel 2003: 36). If we want our student actors to perform at an Olympic level, it is critical that we attend to the pre-theatrical – pre-technique, pre-rehearsal and pre-performance – individual.

Power posing

When working on a production with young actor-athletes, I begin each rehearsal by having them engage in 'power posing' to aid their confidence and psychological resilience. Power influences all aspects of life. Whether it is about resource acquisition in the wild – such as when the alpha dog gets the biggest bone – or about nailing an important audition, feeling powerful simply translates to feeling good about one's self (163–4). Dana Carney, Amy Cuddy and Andy Yapp observed that animals express power through open, space-devouring body shapes and gestures. For example, if a bear or cobra wants to be assertive and powerful, it rises up and outwards as much as possible, thus taking up space in multiple directions. Contrarily, powerlessness is displayed through small, closed shapes and gestures. These researchers examined whether artificially induced expansive shapes could cause an increased positive feeling of power and confidence in humans. They focused on several poses, but the two they cite most are the 'victory pose' in which the arms are outstretched in a wide 'V' and the 'Wonder Woman' pose in which the fists are on hips. Both involve an open, proud chest and a wide, solid stance. They found that power posing – maintaining a bold, expansive posture for a period of time – caused significant physical and psychological changes in the subjects (Carney, Cuddy and Yapp 2010: 1363).

Holding a power pose for just two minutes caused significant changes in the neurochemical make-up of their subjects. These included increased testosterone, a chemical often associated with more aggressive or assertive behaviour, and a decrease in cortisol, the 'stress' hormone. An individual with a high level of testosterone and a high level of cortisol is stressed-out and aggressive (1366), while someone with low levels of both is passive and submissive. If a component of choking is an inability to self-regulate stress and stay focused, then this is a beneficial chemical cocktail to explore. Like Goldilocks's porridge, high testosterone and low cortisol is just right. It allows for confident, cool-under-pressure responsiveness – precisely the kind I was looking for onstage and offstage from Actor X. Simply by taking two minutes out of each class and each rehearsal for power-posing exercises, teachers and directors may be able to simply and effectively induce neurochemical changes in actors that will allow them to make confident, braver choices in their explorations. These poses can help the actor build a cognitive scaffold that allows her to feel and exude a greater sense of calm and confidence.

Besides feeling confident, exuding confidence can be important for the actor-athlete whose professional life will most certainly depend on a great deal of networking and the developing relationships with directors, producers, agents and others. Certainly, power posing turned out to be beneficial for Actor X who was afraid of facing a host of older, more seasoned professionals in a festival environment. In their 2013 paper, 'Upright and Left Out: Posture Moderates the Effects of Social Exclusion on Mood and Threats to Basic Needs', researchers Keith M. Welker and others explain that posture can influence both the perceptions of others and an individual's perceptions of himself. Socially, posture packs a lot of information. The authors point to a 2009 study in which subjects maintained an upright posture or a slumped posture. Those with upright posture were perceived by others as being better job candidates and were projected to be happier as employees by potential employers than their slumped competitors. The same was true for an individual's own feelings of personal positivity and power. Slumping or hunching over, essentially the opposite of a power pose, has equally disempowering effects in self-reported feelings and decision-making capabilities (355–6). In terms of the perceptions of others, the cool, confident individual is more likely to be hired than the individual who maintains a slouched, collapsed shape (Cuddy 2012).

While Cuddy's research on power poses shows that shape and posture can increase one's feelings of power and confidence, posture alone may not create emotional states – for example, upright and proud equals happy, while slouched equals sad. However, power posing does seem to facilitate and allow for more positive emotional experiences. In other words, an upright posture can be a factor in feelings of power and happiness (Kozak, Roberts and Patterson 2014: 415). No doubt any number of social and biological differences might impact confidence. Confounding variables like gender inequality may play a crucial role. Because an individual's psychology is dependent on any number of internal and external variables, one might well assume that the effects of power posing on confidence might be measurably affected by things such as gender. Megan N. Kozak, Tomi-Ann Roberts and Kelsey E. Patterson conducted a gender-based study to see if women would have different experiences with posture than men. Kozak and her team predicted that women, when maintaining a proud, upright posture, would feel more at risk for objectification because 'an upright posture – which involves emphasizing the breasts and physically opening the body to observation and scrutiny – may have induced feelings of self-objectification and thus also made more complex any feelings induced by being in such a posture' (20). The researchers attempted to take into account the complexity of social dynamics contextualizing the internal psychological effects of posture. However, Kozak and her team found that both men and women felt prouder and more confident when maintaining the more powerful, upright posture. Their study does not rule out the robust complexity of social dynamics as confounding variables that an individual has to contend with, nor does it assume that simply holding an upright posture in and of itself is a cure-all to address lack in confidence and create feelings of self-empowerment. But their study does confirm that posture influences mood and confidence.

Power posing, when integrated into a long-term, pre-theatrical and daily practice, can aid the young actor in constructing a personal, cognitive scaffold of confidence that may ward off negative internal and external stimuli that could lead to the actor choking in rehearsal, performance and life. Used in the rehearsal process, power posing helps to create immediate, neurochemical changes and a sense of confidence that the actor can bring to that day's work. While the actor maintains a power pose I also ask him or her to make use of mental/motor imagery, which is the second component of this initial intervention.

Mental imagery

Mental/motor imagery work is a means of internally rehearsing complex staging and choreography. Mental imagery is the act of imagining specific skills or watching them being performed, which results in an improvement in the individual's performance of that task. Employed as an internal, cognitive rehearsal tool, this can be done from either a first-person or third-person perspective. Switching between the two has different benefits. For the Olympic Actor, using mental imagery is a way of rehearsing blocking or other motor sequences in the mind to help deepen their embeddedness in the body. By running through a particularly difficult sequence mentally, the actor is able to examine the motor programme from a number of perspectives, tempos and angles. She is able to slow it down and speed it up, and examine the detail and nuance of the sequence. Additionally, this technique provides an opportunity to imaginatively run through potential problems and chokes, and develop protocols against them. For example, because the role Actor X was playing involved a number of highly specific movement patterns and onstage costume changes, I had her imagine what would happen if one of those things went wrong, then slow down and even pause the images, and figure out in-the-moment, show-specific solutions to the choke.

Mental imagery is a version of the 'mental movie' used in sports psychology. The use of mental imagery has a solid foundation and storied history in acting theory as well. Stanislavski called for the actor to create a 'filmstrip' (*kinolenta*) of every moment of a role as a way to visualize the character's journey moment by moment. This can help the actor to see more possible and specific embodiments for the character in the world of the play.

The actor can also use mental imagery to develop herself. By using mental imagery in rehearsal, the actor can watch staging through her own eyes or as if she were watching, from the third person, her own body as it performs the actions and choreography. In doing so, she can become more confident in the sequences and visualize how to work through troublesome areas.

Sports psychology research has shown mental imagery improves not only mastery of a skill, but also the overall arousal and self-confidence of the athlete. In 'Content, Characteristics, and Function of Mental Imagery', Jean F. Fournier, Sophie Deremaux and Marjorie Bernier cite sources that suggest mastery and confidence are related. Athletes who wanted to improve their efficacy in a skill did so with greater success when they imagined themselves not only performing the skill precisely, but also doing so with self-confidence (735). In 'Imagery Research: An Investigation of Three Issues', Nichola Callow and Ross Roberts explain that, since the early days of mental image research, the prevailing evidence has demonstrated the efficacy of mental imagery in the learning and honing of motor skills, especially in elite athletes. 'Cortical Activity during Motor Execution, Motor Imagery, and Imagery-Based Online Feedback' describes how Kai J. Miller and others used fMRI and PET scans to show that motor imagery – the internal, imagined activation of motor programmes – activates many of the same neocortical areas that actual muscular performance engages. There is significant overlap in primary motor areas when an individual is doing an activity, such as hand or tongue movements in this case, and when he is imagining doing those same activities (Miller et al. 2010). In other words, if Actor X did the blocking in external, physical space or imagined herself doing the actions in an internal, mental space, significant parts of the same neural patterns are activated in the brain. In a related, non-sports psychology experiment, Miller's team asked subjects to move a cursor on a computer screen by using mental imagery. The subjects ran the task three times. By the third time, subjects reported that they stopped using explicit mental imagery – imagining it happening – but instead simply 'thought about the cursor moving up or down to get it to move' (4432).

A mental imagery sequence being rehearsed repeatedly becomes more deeply embodied, and athletes report increased feelings of confidence. I relate this to my experience with young actors, such as Actor X. Employing mental imagery allows actor-athletes to connect how their bodies feel as they move through time and space with how these actions might appear to an audience. Moreover, the conscious use of mental imagery allows performers to begin to create a 'movie in the head' that has the potential to give them more agency in rehearsal and performance.

Even imagining mental imagery as a kind of mental movie requires a degree of mental manipulation, as do all metaphors but, as Fournier's team pointed out, this is not a bad thing. Imagining that the task is malleable – like watching a movie that can be sped forwards, rewound, or paused – may give the athlete a greater feeling of control and, therefore, confidence (Fournier 2008: 737). This movie metaphor turns the actor-athlete into the projectionist: if *I* stream the movie, I can slow it down, start it, and stop it whenever I wish. I have more control over it. The actor-athlete can manipulate the mental image and imagine how the body is moving at specific points in a piece of choreography or take a second to strategize about solutions to a problem. Imagining solutions in slow motion allows the actor to feel more confident about possible choices once everything speeds back up.

One of the leading causes of choking under pressure is rushing into a situation and opting for a potential solution too hastily. Being able to remain calm under pressure and assess the multiple possibilities for action differentiates an elite performer from the novice. Mental imagery is a way to practise cognitive coolness. Beilock suggests that taking the time to assess the problem before jumping into action allows the brain to sift through possibilities and strategies. Giving actors the permission to take time may slow down, in a meditative sense, their engagement with the variety of inputs the brain is juggling. Beilock points to research done by Michelene Chi in the 1980s with undergraduate and PhD physics students who were given the same exam and the same amount of time to complete it. Interestingly, but perhaps unsurprisingly, the more experienced graduate students took more time on each question before answering it. Beilock writes, 'Pausing to assess the situation before starting to solve a difficult problem is one way to ensure success, especially if your first inclination is to look for the quickest and easiest way out' (31). Mental imagery – taking time to visualize possible courses of action – offers a way to do just that. Olympic Actors can take a high-pressure situation or sequence and mentally slow it down, speed it up, rewind it and fast-forward the scene like Netflix for the brain. The confidence that this affords can positively contaminate the actor's performance, giving her a greater feeling of calmness under pressure.

The malleability of mental imagery can also allow the actor-athlete to flip perspectives between first person and third person. In the first person, the actor-athlete has a more concrete imaginative lens. She can see the things she might see as if through her own eyes. According to Lisa K. Libby, Thomas Gilovich and Richard P. Eibach this allows the individual to imagine specific details of actions. Seeing herself from the third person, the actor is able to imagine what the audience sees. This gives her a more holistic, generalized 'big picture' mental image. She can imagine how all of the actors and movements might fit together.

This first-to-third-person shift also has a temporal component. When asked to imagine a socially awkward moment from high school from the first person, subjects remember concrete details such as, 'I used to have problems delivering a punch line, but now I don't.' In the third person, they see a more general continuity over time, for example, 'I used to be a loser, but now I'm the life of the party.' The latter doesn't offer any information about the concrete details of the awkward event, but does offer a holistic categorization (Libby, Gilovich and Eibach 2005: 51–3). The ability to take multiple perspectives on one's self occurs when we organize personal historical narratives. This ability to manipulate perspective can prevent choking under pressure, as evident, for example, in elite athletes' use of techniques that allow them to take different perspectives using mental imagery. As Beilock writes, 'Seeing yourself from multiple perspectives ... can help thwart the spiral of self-doubt and worries that interfere with people's ability to perform at their best' (166).

Taking a third-person perspective allows the actor the chance to imaginatively see what the audience might see. This 'objectification' has a number of benefits, including increased levels of confidence in performance.[1] This may, at first glance, seem counter-intuitive. Even seasoned actor-athletes can often feel a tremendous amount of pressure when they are being watched. Introducing stressors into the rehearsal process can also aid in defending against the dreaded choke. As Beilock suggests, 'Stimulating low levels of stress helps prevent cracking under increased pressure, because people who practice this way learn to remain calm, cool, and collected in the face of whatever comes their way' (34). Using mental imagery to construct positive performances seen in the third person, as the audience might see them, may also introduce some low-level performative pressure. Through frequent exposure, the actor can become less sensitive to the stress. For student actors, part of the problem is that the audience's presence comes late in the game as a new, added uncertainty. Introducing the actor's work with the third-person perspective in the beginning is a way of imaginatively introducing the audience early on as well.

Studies of the preferences of athletes regarding the uses of first- versus third-perspective imagery yield complex results. Research by Callow et al. found that individuals seemed to prefer taking one perspective to another and that they had more vividness and efficacy when taking that preferred perspective. When the athlete preferred a third-person, external perspective, he or she seemed to prefer a particular angle (Callow and Roberts 2010: 327–29). In a study in the 1980s, Kirschenbaum and Wittrock found that successful athletes expressed a preference for a first-person approach, but the study didn't find much evidence for a difference in the benefits of first- versus third-person imagery (90–1). In fact, it would appear that many elite athletes switch perspective depending on the task and time, and benefit from using both points of view. For example, if they are learning a new task far in advance of competition, they use third person. This allows the athlete to observe the action from outside himself. Is the athlete rehearsing a sequence immediately prior to a major competition? It might be best to switch to first person in order to see the action through his own eyes, as it were. Fournier et al. interviewed several professional skydivers who were learning a new routine for a major competition. One elite athlete explained that, when learning a new routine, he tended to see himself from the third-person perspective, as if falling from above. He also described a temporal component in the beginning of his mental rehearsal: he imagined his routines in slow motion so that he could study the nuances of the jumps and mentally execute them in a precise fashion. As he got closer to the actual jump, the athlete used a first-person perspective more as he began owning the routine and developing more nuance, for example, he could pause the image and study hand and arm positions (737–40). In a similar way, actors can employ the switch between first- and third-person perspectives to better engage complex movement sequences or fight choreography.

Mental imagery gives Olympic Actors a tool for controlling and rehearsing blocking and choreography, as well as negotiating potential scenarios they might encounter. A third-person perspective allows for more abstraction and is beneficial when the actor is in a learning phase. A first-person perspective allows the actor to 'see' more concrete and specific details of a particular movement. Additionally, using mental imagery to experience those details from multiple perspectives and tempos affords the individual a greater sense of control and calmness. I engage the actor with both first- and third-person perspectives until he or she begins to recognize their own mental imagery preference. Walking the actor-athlete through manipulation exercises that allow them to play with the tempo and perspective of the image may provide greater mental muscularity, flexibility and control. It seems likely that the more control one feels over image work, the greater the degree of confidence one will feel in physical performance.

The final element in the Olympic Actor project is one that has been long recognized by sports psychologists to have substantial benefit for athletes: positive self-talk.

Self-talk

I partner power posing and mental imagery with positive self-talk as tools for the development of the Olympic Actor. Self-talk can reinforce our beliefs about what we are or, conversely, are not capable of. Fletcher and Sarkar interviewed several of their Olympic athletes, and one medallist eloquently stated, 'If you don't believe that you will win, you'll never win' (674). Confidence has to do with our deep beliefs in our own abilities. Confidence can be reinforced or undermined by how we speak to ourselves and by what language we use with ourselves. The quality of self-talk can be broadly categorized as positive or negative. It is not uncommon to find young actors frequently engaging in negative self-talk that can manifest as self-criticism or self-preoccupation. Dwelling on mistakes, such as a bad audition, and reliving them through a dialogue with oneself can result in anxiety and depression. Positive self-talk, however, can elevate an individual's self-confidence and lead to positive moods (Theodorakis, Hatzigeorgiadis and Chroni 2008: 12).

In a 2008 article entitled 'Self-Talk: It Works, but How? Development and Preliminary Validation of the Functions of Self-Talk Questionnaire', Yannis Theodorakis, Antonis Hatzigeorgiadis and Stiliani Chroni note that positive self-talk has a long history in sports psychology. Self-talk can be defined as 'internal dialogue in which the individuals interpret feelings and perception, regulate and change evaluations and cognitions and give themselves instructions and reinforcements' (12). The goal of self-talk is to influence future behaviour, typically in high-stress situations such as competitions. Research shows that, more often than not, highly successful athletes make use of self-talk more than less successful ones do (10–11).

Positive self-talk has a number of benefits, including increased levels of confidence and positive moods. Theodorakis and his team conducted a study aimed at gauging the effects of self-talk on young athletes and found that all 120 athletes used some form of positive self-talk during training and competition. The reason most frequently cited by the athletes was that self-talk would 'raise/increase/enhance feelings of confidence and belief', as well as 'focus/direct/control attention and concentration for the execution of skills' and 'sustain/increase/maximize effort' (16). All of the athletes also reported something akin to 'I feel stronger' and 'I feel more certain of myself'. Their confidence-belief systems seem to be at least partially contaminated positively by the self-talk. (The researchers did not use any form of pre-questionnaire to ascertain the athletes' base-level confidence or belief in their abilities.) As the researchers suggest, those individuals who are successful at attaining their goals reinforce their hard work with positive self-talk.

Conversely, negative self-talk can have the opposite effect. An actor-athlete who would otherwise perform optimally can undermine her own efforts with negative self-talk. Beilock writes, 'Excessive negative self-talk really hurts your performance. And it won't hurt just your immediate performance. Negativity can also impair performance down the line' (226). Young actor-athletes entering college or conservatory for the first time are likely to be under observation to a greater degree and frequency than they have been in the past. They may not have the natural or learnt ability to ward off negative self-talk that often arises from such scrutiny. Feeling 'not good enough' or judged can lead actors who would otherwise perform optimally to choke. The negative self-talk can impede immediate events like an important audition. Perhaps more importantly, when allowed to build up and become habit over time, negative self-talk can become that chronic voice in the actor-athlete's head that continually thwarts his efforts. A study of competitive gymnasts revealed that those who mentally spoke to themselves in a doubtful manner – 'I hope I don't mess up' – performed less optimally than those who had positive self-talk – 'I can do it' (Weinberg 1984: 146). The old cliché, 'Whether you think you can or not, you're right,' can be rewritten, 'Whether you tell yourself you can or can't, you will or won't.'

By employing positive self-talk early and often, it is possible to counteract both the short-term and long-term effects of negative self-talk. In working with Actor X, for example, we realized that she had a persistent voice in her head that always cast doubt on her efforts. By using self-talk she was able to create a new voice and a new narrative for herself. This does not imply that positive self-talk will rid an actor-athlete of personal, psychological demons, nor is it a cure-all for nervousness and anxiety. In fact, in some cases, a little nervous energy can be beneficial to the Olympic Actor. It is common for an actor to have 'butterflies' in her stomach before opening night. If the actor perceives nervous energy as a negative thing, the related feelings and associations with that anxiety can result in choking during performance. However, if she can employ positive self-talk to steer that energy in a helpful manner, then it is conceivable that she will learn to look forwards to the feeling.

An important aspect of confidence is being able to maintain a sense of ease, especially under pressure. Theodorakis and his team show that positive self-talk has a calming effect on athletes (11–13). The major areas that are reported receiving beneficial effects are attention and focus, motivation and arousal, and confidence and mood. Antonis Hatzigeorgiadis and Stuart J. H. Biddle suggest that pre-competition anxiety is not necessarily a bad thing, but that one needs to understand the direction of that anxiety. In other words, how does the athlete perceive his anxiety? On the one hand, it could be that, through self-talk, the athlete is able to harness the energy of that anxiety and direct it towards the competition. On the other hand, if he perceives it as harmful or negative, then negative outcomes will likely be the result. The researchers found that all of their test subjects experienced feelings of pre-competition anxiety. However, the result of that anxiety experience was mitigated by either positive or negative self-talk. How the athletes talked to themselves about that anxiety caused it to result in more positive or negative performance outcomes (Hatzigeorgiadis 2008: 244–5).

Whatever 'confidence' is, having it is at the very root of the effective and skilful elite athlete. I believe this is also true for actors. My own experience, in the lab and the rehearsal hall, is that actors like Actor X can thwart their own efforts and choke under pressure because of negative self-talk. Self-talk can either feed into anxiety or mitigate it. Self-talk alone will not eliminate pre-theatrical anxiety, whether it is about an audition or a stressful movement sequence. And maybe a little stress is okay.

Conclusion

As I began training and directing student actor-athletes, I realized that few acting methodologies explicitly attended sufficiently (or at all) to the pre-theatrical individual. In other words, few were concerned with preparing the student for the physical and emotional demands of performing under the pressure of observation. Certainly, most techniques make some mention of the actor's need for a strong, supple body and mind. But I soon began to realize that, in order to train and condition confident, functionally fit actor-athletes, I would need to turn to sports psychology and exercise science. In very real terms, my young Actor X was suffering from a lack of self-confidence, which I realized was in direct relationship to her physical well-being. The holistic intertwining of power posing, mental imagery and positive self-talk aided the actor by focusing intention and attention, redirecting her energies to the immediate task at hand and away from external distractions and negative influences. It bolstered her self-confidence, if only for a short period, until she developed a habit of daily, positive self-talk and reflection that was truthful and critical, but also nurturing and supportive. This approach can generate excitement and arousal around the task at hand, whether it is a rehearsal or performance. This pre-rehearsal and pre-performance intervention is certainly not a cure-all. It is intended as part of a larger body of work for improving actor training and performance through the integration of sports psychology, exercise science and cognitive science to produce what I call 'The Olympic Actor'.

5

Becoming Elsewhere:
ArtsCross and the (Re)location of
Performer Cognition

Edward C. Warburton

Introduction

Like most performing artists, professional dancers face a unique challenge inherent in live theatrical presentation: these highly prepared, aesthetically rich enactments must also be experienced as spontaneous, free of artificiality, by performer and spectator alike. How do dancers construct and integrate all the necessary information to perform highly sophisticated physical tasks, lined up in hour-long choreographies that have to be flawlessly remembered, at the same time producing expressions of deep emotional quality that have the power to communicate to others? For contemporary choreographers, the immediate concern is the purpose of invention and the process of selection: that is, the manner in which originality and quality will 'get into' or 'get put into' the dance during the making process. One important issue is how to situate the performer and her potential contributions in the shape of making. For dancers, however, the overriding concern is the memory of, and felt understanding for, the choreography.

When asked about the experience of performing, dancers often speak about how they can be so completely absorbed in the dance that the physical movements literally become automatic. But mindlessly relying on the automaticity of well-practised motor sequences is both risky and unlikely to produce an aesthetically satisfying performance (Chaffin and Logan 2006; Ericsson 2007), whereas thinking too closely about a highly practised skill is a sure way to disrupt it (Beilock and Carr 2001). Dancers must become sufficiently grounded in the choreography such that embodied cognitive mechanisms like perception and memory can contribute to situation-appropriate behaviour, leaving dancers free to respond creatively to moment-by-moment nuances in performance. Performance thus calls into question the difference between 'learning' and 'memorizing', the close relationship between individual and environment, and the influence of this relation on the nature and development of performing bodies' 'situatedness'. Thinking about the thinking behind the doing of dance in this way presupposes an acceptance of dance as cognitive activity. Thus, to understand the ways dancers might reach the necessary state of situatedness for performance, I adopt an 'enactive' approach in the investigation of dance cognition. The concept of enaction has become a cornerstone of the embodied cognition literature, which claims that cognition is 'for action' – that is, the function of the mind is to guide action – and is a 'situated activity' – that is, it takes place in the context of a real-world environment (Varela, Thompson and Rosch 1991).

Because my aim is to connect what has been in the field of dance a largely theoretical analysis of the role of context and situation, I investigate dance cognition in the specific site of *ArtsCross*, a multi-year, international collaboration among choreographers, dancers and academics from Beijing, London and Taipei. From my observations of ArtsCross, I found dramatically different approaches to memorizing versus learning during dance creation and rehearsal. A broad characterization is that most choreographers fell into one of two camps: practice versus deliberate practice (Ericsson and Charness 1994). Where practice in dance often means simple repetition, the deliberate use of indigenous strategies can augment learning in dance in powerful ways. One such strategy that dancers have developed – and ArtsCross choreographers learnt to exploit – is a movement-reduction strategy called 'marking'. In contrast to dancing 'full out', dance marking involves enacting the sequence of movements with curtailed size and energy, by diminishing the size of steps, height of jumps and leaps, and extension of limbs. It can be considered a way of modelling for oneself. This kind of reduction or elimination of particular aspects of the performance during practice is also found in other domains. Athletes, for example, will often go through practice motions without the target object, such as a basketball player practising free throws without a ball before shooting. It is standard activity in theatre to do an 'Italian run-through' – a slang phrase for saying one's lines and moving about the stage extra fast when staging a play to clarify the timing and relative positions of the actors.

In what follows, I unpack these ideas further and argue that learning choreography, as opposed to memorizing movements, requires more than remembering a dance in the context of task-relevant inputs and outputs, where 'perceptual information continues to come in that affects processing, and motor activity is executed that affects the environment in task-relevant ways' (Wilson 2002: 626). Where situated cognition emphasizes the importance of real-time context in establishing meaningful linkages with learner experience and in promoting connections among knowledge, skill and experience (Lave 1991), I contend that what actually performs as well (or better) for ArtsCross participants may be an abstracted, 'off-line', dance-marking strategy. This strategy decouples dance from specific environments and physical inputs and outputs while it simulates aspects of the performance and assists in learning and enacting it. I suggest that this is particularly the case when the complex relationship between the object of the dance and the subject who dances is explored in a multicultural, multilingual setting such as ArtsCross, which occurs far from home for the majority of the participants.

Dance enaction

As I have argued elsewhere, to be trained as a dancer today is to be enculturated into a world of meanings and movements (Warburton 2011). Despite a recent proliferation of dance studies, exploring social, cultural, historical and philosophical aspects of dance, there has been little effort to advance a psychology of dancing. Some investigators consider dance as 'thought made visible', but fail to provide an adequate domain-specific account of how the candidate cognitive processes that underpin dancing differ from gesture communication or musical or athletic performance (Stevens and McKechnie 2005). Other accounts of the mind and dance assume that experiences of dancing, like other human behaviours and mental events, are entailed or instantiated by either physical processes in the brain–body or cultural processes in society (e.g. Bläsing, Puttke and 2010; Desmond 1997; Farnell 1999). In principle, these assumptions must be correct, since dance can be explained by events in the physical or social world. But materialist accounts often go one step further by assuming that experiences can be redefined as nothing but these causes, and therefore must be understood solely in terms of them.

An adequate account of dance experience requires more than general mental processes or specification of causes; it requires a description of the content (i.e. what is felt/thought) that is common to all experiences of dance and that distinguishes one experience from another. Content-rich concepts emerge at the level of psychological description and are causally constituted by neurobiological processes. A concrete example of a content-rich concept is the dance-specific tool known as 'marking' mentioned previously. Literally 'to mark time', it is a memory device that dancers employ to mark particular moments in the dance – compressing their movements in space and chunking sequences in time – in order to commit to memory long passages of choreography. (Choreographers understand this need and may tell dancers not to dance 'full out' but to 'mark it'.) Marking is part of the practice of dance, pervasive in all phases of creation, rehearsal and reflection.

Virtually all English-speaking dancers know the term, though few scholarly articles exist that describe it.

This perspective suggests a method of inquiry that generates insight into dance experience by taking the enactive approach. Enaction is an especially powerful theoretical lens for understanding dance cognition for three main reasons. First, it posits a mental model that encompasses three intertwined modes of bodily activity – modes that resonate with three intertwined realms of dance experience: self-regulation (somatic realm), sensorimotor coupling (kinaesthetic realm) and intersubjective interaction (mimetic realm) (Cohen 1993; Kimmerle and Côté-Laurence 2003; Press 2002). Second, the enactive approach emphasizes the roles of emotional and relational experience in meaning-making. Dancing as a form of 'keeping together in time' highlights one of the many ways emotion and cognition are linked from early perception to higher-order reasoning (Phelps 2006; McNeill 1997).

Third, the explanatory power of enaction lies in the ways it treats distinct claims in approaches to embodied cognition. For example, one can distinguish online (time-pressured) aspects of embodied cognition from offline (mental simulation) aspects (Wilson 2002). Traditionally, the various branches of cognitive science viewed the mind as an abstract information processor whose connections to pressures of the outside world are of little theoretical importance. On the one hand, the ideal cognizer steps back, observes, assesses, plans and takes action (or not), and then reflects on what happened. Phenomenological accounts of 'lived' experience, on the other hand, tend to view cognition as primarily in-the-moment and online; by definition, the situated being is an ecologically oriented one who is in near-constant interaction with the things that the cognitive activity is about (Gibson 1979). Because the enactive approach views knowledge as constructed in action through emergent and self-organizing processes, it can account for the workings of both online and offline cognitive (and emotional) processes simultaneously. It is this latter claim that motivates the present inquiry.

I am particularly interested in the ways that a movement-reduction strategy in dance, such as marking, reflects the human capacity for the shaping of tools to match a mental template; the most commonly cited of these are language, allowing communication about hypotheticals, past events, and other non-immediate situations, and depictive art, showing the ability to mentally represent what is not present, and to engage in representation for representation's sake rather than for any situated functionality. These species-defining features of human cognition reflect the sometimes offline, context-independent nature of human thought and action that can operate concurrently with real-time, online, situated cognition as a fundamental principle of human cognitive architecture.

Why ArtsCross?

ArtsCross is a multi-year, three-way collaboration among artists in London, Beijing and Taipei with a double focus. On the one hand, nine choreographers from the three cities were selected to work over three weeks with mixed groups of dancers, also from the three cities, to create a ten-minute work on a theme. The three editions included Taipei 2011, 'Uncertain ... waiting'; Beijing 2012 'Light and water'; and London 2013 'Leaving home: being elsewhere'. In each of three editions, a majority of the dancers and choreographers travelled to participate, with a handful of dancers participating in all three editions. On the other hand, a cluster of academics, most but not all from these cities, gathered to watch, to reflect upon and to exchange ideas about the process in action. In this setting, the purposeful striving towards 'something' arises from the requirement to resolve the problem of invention in three weeks. This given condition represents a uniquely teleological view of creativity, which in a broad sense means that the ends govern the means: that 'something' must be directed towards a particular end to generate creative outcomes.

The belief implicit in this method is that invention depends not on special mental processes but on special purposes. That is, rather than a churning unconscious or sudden illumination, what makes creating special is not so much its component mental processes but their organization and direction, and that organization and direction derive from an end in view, however broadly characterized or vaguely grasped. As suggested by the founding director of the project, Christopher Bannerman, one of the goals of ArtsCross was to invite participants to meet the challenge of an 'aesthetics of elsewhere' as a multilayered process – artistic, cultural, political, social experiment – with public performances as outcome. (It is important to note that the academics were required to blog daily and write a paper for post-performance academic conferences that concluded each of the three editions.) Along with the positioning of 'elsewhere' at the heart of the matter, the interrelationship between process and product is one that was open to debate. But ArtsCross deliberately positioned the performance as a result of cultural exchange and artistic process, combining 'the investigations of creative artists who are active in the arts marketplace with the debates of academic observers for whom performance was a practical as well as an intellectual endeavor' (Bannerman, personal communication).

In this context, all participants assumed a teleological stance, but that did not mean that one knew at some level just what the product would be before making it, or that motivated effort was a substitute for creative process. Participants understood that necessary deadlines were insufficient for generating good ideas. Making means selecting. And yet intents to create or to satisfy unreasonable demands, or both, can pattern and bias those component processes towards creative accomplishment. Purpose shapes process. Indeed, it was the very process of selecting from an infinity of possibilities those properties that go beyond what a person can simply, straightforwardly and effortlessly do that was the underlying premise of this creative and cultural enterprise.

All of the ArtsCross choreographers confronted this tension directly, as did their dancers. As designed and implemented, ArtsCross represents an extreme example of the already changeable, diverse, peripatetic and time-pressured set of practices governing the field of contemporary dance today. It was also (in part) an ideal setting for investigating the nature of cognitive 'situatedness'. What occurs when dance making takes place in a physical location and teleological context that is largely foreign to a majority of participants? More than making a dance, the problem for many choreographers became, how do I 're-locate' these dancers far from home in order to achieve something more than a satisfactory outcome?

What performs?

For viewers, what performs in dance is often the memory of seeing it. For dancers, as suggested above, the nature of participation in the creative process is paramount. In considering ArtsCross from a dance enaction perspective, the difference between 'learning' and 'memorizing' a new dance is ultimately what performs onstage.

Many choreographers and instructors tend to view dance content as something that should remain implicit and physically embedded in movement, with understanding and ability emerging spontaneously via repetition (Warburton 2004). Here, learning and memorization are one and the same, making its magic by simple association. The assumption is that moving precedes thinking: this implies a one-way, hierarchical model of learning in which high-level cognitive (e.g. analysis), emotional (e.g. relating) and physical (e.g. jumping) operations grow out of low-level ones (e.g. comprehension, empathy, strong feet) (Zohar, Degani and Vaaknin 2001). For instance, rather than imagining or thinking through the motivation for a particular role, this perspective understands the emotional range required for that role as developing out of a kind of motor resonance for the movement itself, a _feeling for_ the movement, which provides the dancer with all the emotional capacity and choices needed to express fully the character's intent and relationships. By practising the movement over and over again, high-level emotional operations (relating) will grow out of low-level ones (empathy), or so the reasoning goes. This view is particularly prevalent in dance where folk beliefs about dance as being somehow non-cognitive and non-deliberative remain intact.

For dance performers, one of the most relevant questions about rehearsal pedagogy is quite simply whether the memory they attained in the studio will be reliable on stage. In a study of musicians, Chaffin, Logan and Begosh make the case that different types of memories are learnt in different ways for different purposes and have different properties:

> The memories that develop spontaneously while learning a new piece take the form of *associative chains* in which each passage cues the memory of what comes next. Deliberate memorization transforms the motor and auditory chains, making them *content addressable*. (2008: 352)

Pure practice or rehearsal by rote is very good at building associative chains. Do this, then this, then this ... repeat. Most dancers and dance instructors are accustomed to relying heavily on repetition to solidify skills and help recall choreographic structure. The approach relies on context-specific knowledge and behaviour that can be accessed easily by starting from the beginning every time. It is especially conducive to time-pressured situations. Associative chaining performs very well because it is concrete and inherently situating: what you are dancing at the moment reminds you of what comes next. For some ArtsCross choreographers, it also circumvented the need to confront any significant differences among dancers since it located them in a comfortable, predictable, reflexive place. Repetition drove rehearsal. Instead of dialogue or joint construction or collaborative problem-solving, 'Do it again, from the beginning' is a more common refrain in the dance studio. This approach reflects a widely inherited dance practice, stemming in large part from a tradition of choreographic creation and dance pedagogy that values associative chaining and rehearsal by rote.

This is not to say that repetitive practice methods always disappoint in final performance. I found choreographies specifically built on progressive movement patterns and pedagogies of repetition to be oddly compelling. One performance in particular, Su Wei-Chia's *Free Steps* (London 2013), aimed for a finished look but eschewed any sense of beginning and endings and arcs in between. Over the course of ten minutes, *Free Steps* starts as it continues as it ends: five women clustered in a pool of light, frond-like arms wafting through the air, heads and torsos and legs slowly, deliberately turning, folding and twining, feet inching the group along a diagonal pathway while ambient sound immerses the space. While it is likely that his dancers mentally practised outside of rehearsal, Su Wei-Chia disallowed any forms of practice (such as 'marking') other than linear repetition during rehearsal. On stage, the overall effect reproduced the vacuity of the rehearsal room. As one critic put it, *Free Steps* is 'a meditative, mind-emptying experience'.

Experienced performers, however, appreciate that associative chaining of movements is just the first step in learning new work. Associative chains are more likely to be implicit and involve procedural knowledge. Content-addressable memories are more likely to be explicit and involve more abstract, declarative knowledge. Much work is needed to create reliable, content-addressable memories that address the range of qualities and artistic intentions of a work (Chaffin, Logan and Begosh 2008). For example, prior to my observations in the first edition, I wondered, how would ArtsCross dancers possibly integrate in three weeks all the necessary information that would allow them to perform passably these highly sophisticated creative works? All the choreographies were works in progress. Dancers were not remembering restaged choreographies, using previously acquired movement techniques and supported by a coterie of répétiteurs and rehearsal assistants. For most, the new movement vocabularies presented a steep learning curve. Instruction was in a second language that often required translation. Moreover, many dancers were expected to learn more than one work.

Dance marking

In thinking about questions of making and learning – and of the performers' predicament in achieving a highly practised, seemingly spontaneous nonchalance – I considered what kinds of strategies dancers might employ to create reliable, content-addressable memories. What I found was the widespread usage of dance marking. These dancers performed an attenuated version of the choreography during rehearsal, enacting the sequence of movements with curtailed size and energy; for example, by diminishing the size of steps, height of jumps and leaps, and extension of limbs. I observed that marking was happening spontaneously in very different ways and for very different purposes throughout the ArtsCross rehearsal process in each of the three editions (Taipei, Beijing, London), seemingly as part of dancers' ongoing efforts to grasp various aspects of the new and alien choreographies.

Following observations of the Taipei and Beijing editions, I collaborated with colleagues in dance and psychology to do experimental research on dance marking and the cognitive benefits of movement reduction. Several questions of interest arise from this phenomenon. One set of questions involves the purposes of such reduction. In some cases, such as whispering, the purpose is to alter communicative functions (e.g. to include some listeners/watchers while excluding others, or to create a conspiratorial, humorous or ironic tone). In other cases, such as sub-vocalization or reduced gesturing or pointing movements, the purpose may be purely internal, to aid the cognition of the individual without regard to others. Marking in dance is particularly interesting in this regard because it is widely assumed in dance that the purpose of marking is simply to conserve energy. But elite-level dance is not only physically demanding, it's cognitively demanding as well, requiring one to concentrate on many aspects of the desired performance: from the most basic elements of accurate body positioning and correct timing, through higher-level chunks of choreographic phrasing, to more subtle features of performance expression in specific ways at specific points in the choreography. A dancer who is rehearsing must concentrate not only on all of these features, but also on the physical demands of the movements, such as maintaining balance, which may also impose a cognitive load. Thus, marking may serve not only to conserve physical energy but may also relieve cognitive load. In fact, many dancers develop marking systems that are highly representational, rather than just miniaturized performance, such as using a finger movement to represent a turn while not actually turning the whole body. These kinds of strategies may allow them to rehearse some aspects of the performance (e.g. the correct timing, or the use of the head and the arms to convey emotion) while not needing to allocate attention to other aspects (e.g. maintaining balance during a turn, or reorienting oneself in space after turning).

In our study, we tested the *embodied-cognitive-load hypothesis* of dance marking by manipulating whether dancers rehearsed by marking or by dancing 'full out'. If marking serves a purely physical function, then no difference in performance should be observed, or even possibly there should be an advantage for the 'full-out' condition, since rehearsal more closely matches tested performance. In contrast, the cognitive-load hypothesis predicts better performance on a dance routine that was rehearsed with marking, due to reduction of cognitive load allowing more effective memory encoding. Subjects were advanced ballet students with a mean of 14.4 years of ballet training and 16.0 total years of dance training. Each subject served as his/her own control due to the within-subjects design. In addition, subjects were assigned to two groups matched as closely as possible. Two 64-count ballet sequences were choreographed (Routine A and Routine B) and both performed to the same piece of music. The two sequences each consisted of eight movements. Each movement within each sequence was then assigned a movement quality with which it was to be performed. The routines were designed to be fairly easy for advanced dancers to memorize and perform, with the intention of placing the burden on integrating the movement qualities rather than ballet steps. Qualities provide a rich problem for a dancer to commit to memory, which elite dancers will continue to work on long after rapid consolidation of the sequence of steps. A multi-day procedure was chosen to take advantage of the role of sleep in memory consolidation. Each movement of each performance was scored using an analytical scoring method based on Laban Movement Analysis, which assessed whether each quality had the appropriate features of weight, time and spatial intention. Each feature of each movement was scored as a 0 or a 1, for a total possible 48 points for each routine.

Our research group found that loosely practising a routine by 'going through the motions' improves the quality of the final dance performance by reducing the mental strain needed to perfect the movements. By decreasing demands on complex control of the body, marking reduces the multilayered cognitive load used when learning choreography (Warburton et al. 2013). A fundamental insight of the embodied cognition literature is that even very abstract thinking may be parasitic on evolutionarily older brain systems that originally subserved purely sensory and motor interactions with the world. Our research studied an overlooked aspect of this legacy: that resources allocated to controlling the body, at least at a high level of complexity that pushes expert performance to its limits, are inseparable from resources needed for thinking at a more abstract level about the material to be memorized. Inevitably, resource competition will occur. Our finding suggests that dancers have in fact developed a tool that confers processing benefits during the rehearsal process.

One set of questions that our research did not address involves the ways in which the reduction occurs, and how a movement can be altered and still be considered the 'same' movement. For example, when signed languages are whispered, the signs may be displaced to a very low location in the signer's signing space, disrupting a supposedly defining feature – hand location – of many signs. Yet other fluent signers readily understand this whispering. As a different example, marking in dance serves no communicative function and is rarely formally taught, and so each dancer may establish her own 'vocabulary' of reduced movements to represent the fully formed movements. Our study examined American ballet dancers learning classical choreography in a controlled environment. How might marking be used by a diverse group of contemporary choreographers and dancers trained in varied ways, working across languages and techniques and creating original works under severe time limitations? To what degree is marking a *context-independent* strategy for dancers seeking to generate reliable, content-addressable memories? What does marking look like 'in the wild', so to speak (Hutchins 1995)?

No Lander

For the London edition, I focused particularly on how ArtsCross choreographers located their creative process in the body-minds of dancers and the devices they used for relocating performer cognition in a strange, though recognizable, land. When one has only three weeks to create and perform, what method(s) can drive such a creative state of becoming elsewhere? More specifically, what happens when marking is used strategically by choreographers working with a diverse group of dancers – from different cities with different trainings using different languages – to enhance memory and integration of multiple aspects of a piece precisely at those times when dancers are working to master the most demanding material?

I once again observed dance marking happening everywhere, in very different ways and for very different purposes. In the rehearsals of London-based Italian choreographer Riccardo Buscarini, however, I observed marking as a daily, explicit practice. His choreography for *No Lander* opens on a solitary seeker, who is taken up by a group of four men, and together they lean and pitch like the prow of a vessel pushing forwards. Then they go into reverse, blocking and impeding each other's movements until the group breaks up, leaving its dancers stranded.

When I spoke with Buscarini about the use of marking he mentioned using it in a multifaceted way: tracing the spatial trajectories, sensing one's contribution to the visual design, developing a common relationship, becoming a connected ensemble. He said,

> There's a training of listening (to) one another and also (the) group (connecting) to the piece … yesterday I used this image and I think it was brilliant because they went into marking in the most amazing way. I asked them to imagine that the piece was already drawn in the space and they just had to fit into each configuration [laughs] as if the whole thing was just constructed and easy … you just go into it and out of it. … I use a lot the images of traces … every thing that I do is based on leaving traces in the space. So yesterday, I asked them: remember the piece as a drawing, a system of traces in space, and mark through the piece as if you just hitting those positions, just going through a path that is already drawn. And that helped them so much, and I think it looked so crystal clear from the very beginning. (Buscarini, ArtsCross: London, 7 August 2013)

For Buscarini, marking was understood as much more than simply a way to get on the same count, save energy, or as a form of physically thinking through the dance. His objective was to get the dancers to mentally project more detailed structure onto the architecture and poetry of his work. By setting up images of searching against images of blindness and requiring mental projection into the future, Buscarini's creation and rehearsal pedagogy seemed to reflect our unique human capacity for being here now and simultaneously becoming elsewhere. In performance terms, it is the imperative to embody fully the moment of movement even as one envisions next steps. As one critic told me after the show, 'It's a simply constructed work that swells with poetic significance.'

Interestingly, in addition to the changes in creative expression, I also observed a subtle change in interpersonal relations from the first to last rehearsals of *No Lander*. In the beginning, ArtsCross dancers in every rehearsal in every studio would separate at each rehearsal break to check their mobiles and text messages to friends and family. The halls would be filled with silent dancers tapping out electronic updates. In contrast to other choreographers' rehearsals, the individual and collective attitudes and communication patterns of Buscarini's dancers appeared to shift rapidly. Buscarini specifically spoke with them about 'marking as a group'. He later reflected with me about its use in connecting dancers to create an articulate ensemble. By the end of the first week, Buscarini's dancers were unique in that they began spending rest breaks together, laughing, gesturing and showing off for one another. I did not observe this shift in other choreographers' rehearsals.

Though there were obviously many variables at play, this unexpected observation raises the possibility that marking may serve differential purposes for individual versus group cognition and behaviour. Buscarini's use of marking may have had a role in generating a satisfying interpersonal contact without the need for linguistic translation. Dance marking may have unconsciously brought dancers physically and mentally closer as they struggled to solve problems of dance creation, language barrier, cultural dislocation and stylistic diversity. Asking dancers to do this kind of group marking exercise evidently requires a kind of joint attention and communication that one does not find in marking dance solo. Future research on the nature and use of dance marking in groups might profit from *No Lander*'s example.

Concluding thoughts

Dance marking at ArtsCross can be understood as one example of the ways choreographers and dancers probed the complex relationship between the object of the dance and the subject who dances. Many ArtsCross choreographers used images and ideas to meet head on the dancing-subject to dance–object relationship and to partially dissolve cultural and linguistic challenges. For instance, choreographer Tung I-Fen explained to me that, in her *Sound of Numbers*, she used the abstract idea of numbers as a context-dependent strategy for cutting across language barriers. At one point, the dancers all talk at once; from this unintelligible babble, they begin to recite numbers and the piece then develops arithmetically: movement added to movement to equal phrases; accumulations of dancers to build up groups; divisions and subtractions. The context of the work depends, and comments, upon the presumed lack of comprehension in the face of our human propensity for counting (and moving). It extends this confusion to the audience. Cold, but effective.

Buscarini, however, relied on a context-independent movement-reduction technique in the form of dance marking to bring dancers together: an embodied but essentially abstract strategy for learning choreography. Far from being a necessary evil in the rehearsal process – that is, to save energy – marking can enhance dancers' cognition of movement in time and space with richer aesthetic schemas. Even as expert dancers locate themselves in movement by building chains of associative procedural memories, marking builds content-addressable memories that allow them to anticipate potentially expressive moments in the dance: to respond creatively to the fleeting, liminal, 'becoming elsewhere', placed somewhere between here-now and there-next.

In ArtsCross, dance marking as shared conceptual device and rehearsal pedagogy can thus be theorized as a kind of trans-lingual (non-verbal) practice for bridging the incommensurability of language so as to establish and maintain a hypothetical equivalence between movements and their meanings (Chiel and Beer 1997; Port and Van Gelder 1995). In this way, the choreographer's use of dance marking as both a *context-independent* strategy and *situating* activity became a highly effective way to cross languages and cultures (Clark and Chalmers 1998). Marking as cognitive mechanism ('what performs') acted as a kind of physical re-languaging that relocated dancers who were far from home in order to create a shared physicality and a joint, felt experience of the work in progress.

Like most theatrical performance, dancing emerges from a continuous stream of evolving affect, conceptual processing, physical sensation and psychomotor skills all bound together in time and space to create connections between individuals and ideas. The conceptualization of thinking in dance as *dance enaction* not only opens up dance to promising areas of cognitive research, it also leads to some otherwise unattainable conclusions about the phenomenon of dance cognition and performance.

On the one hand, dance enaction enlarges phenomenological inquiry in that it probes the complex relationship between the object of the dance and the subject who dances. Rather than cordoning off subject from object in a traditional phenomenological manner, one finds the physicality of marking is experienced as an intrinsic part of the dancing experience of the moving subject as she develops understanding of the dance. However, dance enaction further problematizes the concept of embodiment for cognitive scientists more interested in cognitive processing capacity than meaning-making. The activity of dance marking not only allows for subjectivity to be accessible through the perceptual appearance of physical body 'movement reductions', but it also can account for the workings of both online and offline cognitive processes simultaneously. Research on dance enaction thus opens a new window into the processes by which performers build up an adequate representation of a performance sequence that is complex both 'horizontally' (extended over time) and 'vertically' (multilayered at any given moment).

Training, Insight and Intuition in Creative Flow

Christopher J. Jackman

Introduction: Ignorance as bliss

In Heinrich Von Kleist's dialogic fiction, 'On the Marionette Theatre', our narrator describes meeting with an old friend, Herr C-, who was appointed principal dancer at the local theatre. With perfunctory speed, this friend turns the conversation towards what he recognizes as remarkable grace displayed by pendulum marionettes 'dancing' in the marketplace. The friend lauds the puppet's ability to perform without the self-conscious affectations of human dancers, but as translator Idris Parry notes in the preface to his translation, this piece is not so much a paean to puppet construction as an attempt to grapple with the stilting self-awareness at the root of those affectations, or the sense of distance that human consciousness establishes between us and actions in the world (1981: 10). In one striking example, the narrator recalls a young man at the bathhouse; this young man unthinkingly struck a graceful pose, but was unable to replicate that natural expression of physical grace once it had been brought to his attention. It appeared that 'an invisible and incomprehensible power, like an iron net, seemed to spread over the free play of his gestures' (Von Kleist [1810] 1991: 239). His natural ease was calcified by a touch of self-awareness.

Von Kleist's account is swathed in a theocentric epistemology, implicitly mythologizing the Fall of Man as a recurring event rather than a historical one: knowledge drives us further from the unselfconsciousness of a suspiciously utopian 'natural state' (Parry 1981). Nevertheless, Von Kleist's observations insightfully describe an all-too-familiar experience for any novice actor, whose self-awareness seems to hobble her capacity to 'act naturally' as well as her ability to 'be in the moment' when performing a more sophisticated score of dramatic actions.

Similar phenomena have also been observed in studies of performance in sport and music. Studies of professional athletes by Beilock and colleagues (Beilock and Carr 2001; Beilock et al. 2002) found that focusing one's attention on a familiar task significantly degraded the accuracy and efficiency of the act. Music psychologist Frederick Seddon identified a mirrored trend in his study of improvisational jazz performance, noting that when the individual is least self-absorbed, she is best equipped for creative 'risk-taking and self-challenge' (Seddon 2005: 49). In each case, self-consciousness has deleterious effects upon one's capacity to perform spontaneously, casting the body as an estranged object of attention instead of a coherent medium for action. It is a bit like a child who has taken apart his calculator to see how it works: it stops functioning once the pieces are pulled apart, and now it is unclear how they ever functioned together in the first place.

While we may not be able to return to innocence and un-disassemble the calculator, as it were, there remains the possibility that knowledge could help us put it all back together again. Indeed, Von Kleist suggests we may surmount the difficulties created by self-consciousness in performance by perfecting our knowledge and skill. Functional grace re-emerges 'when knowledge has, as it were, gone through an infinity; thus, grace emerges most purely in that human form which either has no consciousness at all or an infinite one' ([1810] 1991: 240). Optimally unaffected performance thus lies in the absence or resplendence of knowledge, 'in a puppet or in a god' (Ibid). This chapter will follow up on Von Kleist's recommendation by considering how continued training may help the performer become, if not a god, then at least an expert, capable of performing skilfully, spontaneously and unselfconsciously. It will describe the role of self-consciousness in training and performance, and will employ an enactive model of cognition to discuss how training may be optimized to support increasingly creative applications of psychophysical skills.

Becoming the spontaneous strategist

Elizabeth Bridges's reading of 'Marionette Theatre' attributes the prescience of Von Kleist's account to his astute regard for the embodiment of mind. Rather than relegate 'the body' and its antecedent functions to some crude, lesser status, Von Kleist regards it as a pre-eminent source of invention and grace, finding beauty in the economically efficient response of natural forms to physical forces. The gestures of the thoughtless puppet or the instinctual animal are a graceful, economical and unaffected reflection of divine physics, while the innocent or animal mind is a responsive extension of that body. Limbs and brains alike express nothing less than natural law, they function 'at a higher level of complexity' than the simple mind can grasp (Bridges 2012: 81–2), and work beautifully until thwarted by our inept intrusions. Von

Kleist hereby establishes his own nascent brand of cognitive materialism, which 'implicitly engages in an inquiry regarding the nature of human consciousness as an object of scientific study and everyday experience, ... as natural rather than supernatural' ([1810] 1991: 76).

Bridges roots the utility of Von Kleist's analogy in its potential to both articulate and problematize our phenomenological experience of mind-body duality, treating the marionette 'as an arrow pointing at other possibilities of seeing the body and its relationship to consciousness' (Ibid: 88). However, apparent gaps in Von Kleist's analogy prove especially revealing, as the reader is left to identify a reasonable counterpart to his small-stage puppeteer. Von Kleist has minimized any distinctions between the mindless puppet and mind-full god, locating spontaneous grace in the heuristic reactivity of the puppet to the string, but even if an 'infinitely knowledgeable' performer could recapture that reactivity, the reader must ask: What force now impels her to react? Is the performer at least partially responsible for 'pulling her own strings', and if so, what form of knowledge or training prevents the intrusion of consciousness upon the inception of impulse?

As pedagogue Clive Barker argues, a sufficient mastery of the kinaesthetic principles underlying instrumental performance is essential for theatrical improvisation, as with jazz music, but even if technique 'enable[s] a performer to embark on wild adventures without preconceived intention and reflective considerations' (1997: 13), rote replication cannot account for the performer's capacity to recognize a worthy adventure. Von Kleist's conception of knowledge and technical mastery thereby fails to account for how practitioners become better at redeploying their technique in novel contexts, at least prior to reaching the utopian terminus of an infinite consciousness. In order to describe the performer as a potentially self-directed, creative agent, we must therefore consider how training alters both the quality of one's performance and the quality one's consciousness in that performance.

For his part, performance scholar Phillip Zarrilli suggests that training benefits the practitioner by affording an implicit awareness of and control over the subtler psychophysical dynamics underlying her practice (2008: 81). Combative sparring, for instance, rarely employs the advanced ornamental forms or rigid kata sequences the fighter rehearses in when training, but those systems of practice foster an attenuated sensitivity to the fighter's place in her combative environment. Sociologist-cum-boxer Loïc Wacquant echoes Zarrilli's account with his own:

> Training teaches the movements ... but it also inculcates in a practical manner the schemata that allow one to better differentiate, distinguish, evaluate, and eventually reproduce these movements. It sets into

motion a dialectic of corporeal mastery and visual mastery. ... Every new gesture thus apprehended-comprehended becomes in turn the support, the materials, the tool that makes possible the discovery and thence the assimilation of the next. (Wacquant, quoted in Lizardo 2009: 720)

Eventually, the trained body becomes 'the spontaneous strategist; it knows, understands, judges, and reacts all at once' (Wacquant 2007: 155).

This account becomes admittedly more complicated, but no less germane, as we attempt to delineate what counts as training for the diverse practices we might variously describe as 'acting'. First, we would be hard-pressed to identify a discrete toolkit of so-called 'acting skills', for, as Lutterbie notes in his *Toward a General Theory of Acting*, actors' efficacy depends at least in part on their ability to 'use the tools they have in common with everyone else' (2011: 103).[1] Second, technical skills may be refined in specialized psychophysical pedagogies[2] or adopted with varying degrees of formality in local studio practice, but rigorously moderated pedagogies are the exception rather than the rule in contemporary Western actor training, where a multiplicity of mongrel pedigrees typically intersect in any given school, on any given stage or in any given history of individual practice (see Renaud 2010).

Ultimately, the intractable *techne* of acting, insofar as one might exist, extends to encompass the *techne* of the actor's self – the active social intelligence that is shared with one's audience and culture – and belies accounting as a discrete technical skill set. This returns us to Von Kleist's foundational question of consciousness, since *what* it is we perform in training ultimately informs *how* it is that we perform. To quote Barba at length:

> Training does not teach how to act, how to be clever; it does not prepare one for creation. Training is a process of self-definition, a process of self-discipline which manifests itself indissolubly through physical actions.
>
> This imperceptible daily transformation of one's way of seeing, approaching and judging the problems of one's own existence and that of others, this sifting of one's own prejudices and doubts ... is reflected in the work which finds new justifications, new reactions: thus one's North is displaced. (Barba 1999: 79)

Taking both Barba and Von Kleist on board, a comprehensive understanding of psychophysical training must therefore account for how skill acquisition and mastery inform cognitive processing.

Autonomy in cognitive processing

Cognitive scientist Keith Stanovich suggests that deleterious self-consciousness may be rooted in the functional distinctions between intuitively led behaviour and analytic thought. He proposes that the majority of our cognitive function is negotiated through these latter processes, distinguished as Type 1 processing.[3] Type 1 processing is autonomous, encompassing innate procedures like behavioural and emotional regulation, implicit learning processes and the 'firing' of overlearned association (Stanovich, West and Toplak 2011: 104). Multiple Type 1 processes run at any given time; it is quick, computationally cheap and is our default mode of cognitive practice. When Lutterbie notes that 'when we navigate a crowded street, we do not distinguish between thought and movement but think in movement' (2006: 156), we may take this thought-in-movement as an example of Type 1 processing at work.

Type 2 processing encompasses what we might refer to as our 'higher' cognitive processes, which include analytic thought, critical reasoning and conscious control. While Type 2 processing supports complex and abstract thought, these processes tend to operate serially, performing one task at a time in a manner that is comparatively slow and more methodical. Type 2 processing requires far more effort and energy, and since from an evolutionary perspective we tend to be conservative with our energy expenditure, heuristic responses typically run their course (Stanovich ,West and Toplak 2011: 105; Hogarth 2001: 194), to such an extent that it is most often 'responsible for building a narratively coherent description of the behavior engaged in by the individual', justifying our behaviour as a set of self-motivated intentions 'even though it did not initiate much of it' (Stanovich 2004: 49).[4] Consequently, when an action is brought under Type 2 control, it lacks the heuristic efficiency with which we might otherwise perform it, bringing with it the 'iron net' of self-consciousness and methodical inhibition that plagued Von Kleist's young man in the bathhouse.

One's attention to details, or consciousness of mistakes and imbalances, imbues early training with 'a painful sense of expropriation of [one's] spontaneity' (Barba 2002: 101), so actions are slow and cognitively effortful. But insofar as the performer's objective is to act effectively and efficiently in a manner that suggests natural grace, the poison of self-consciousness is its own cure. Type 2 processing may be employed to structure disciplined psychophysical training through explicit learning, as deliberative shifts of volitional attention allow one to focus on the mechanical components of novel skills and to correct maladaptive forms. A toddler's fitful attempts at learning to walk belies the ease with which these new skills are deployed in Lutterbie's example above, but an awareness of imbalances and insecure footing provides the basis for improvement.[5] Type 2 processing may also steer our attention towards instructions being provided by a teacher, or to our memory of those instructions, or perhaps towards the body itself, as when a pupil checks to see if his heels are in proscribed alignment for the lion pose in *kalarippayattu* (Zarrilli 2008: 54). Type 2 processing thus plays a key role in structuring learning so that the performance of novel skills will no longer require overt management.

It takes some time until a novel skill is sufficiently familiar to be managed by autonomous Type 1 processing, and even longer until this processing may be sustained without error, but this ideally eliminates the need for self-conscious awareness in the actor's physical execution of her task, which might otherwise divide an impulse from its expression. This phenomenon recalls the removal of expressive blocks as advocated in Grotowski's *via negativa*. According to Grotowski's own example of voice training, the vocal apparatus must be 'infinitely more developed than that of that man in the street', so that it is 'able to produce sound reflexes so quickly that thought – which would remove all spontaneity – has no time to intervene' (1968: 35). Or, as Von Kleist concludes, we must 'eat once more of the tree of knowledge in order to fall back into the state of innocence' ([1810] 1991: 240), availing of cognition that is at once graceful and heuristically managed.

Enactivist model of skill in sense-making

Any analysis of embodied skill will be necessarily fraught, since implicit skill is not really 'known' according to logical or semantic structures. Even when Type 2 processing has structured our training and thus the circumstances of implicit learning, it does not follow that we will or can always have explicit theories encapsulating the procedural knowledge we acquire (i.e. we may learn to catch a ball without being able to describe the full functioning of our musculature or articulate a comprehensive theory of gravity) (see Hogarth 2001: 106–7). Experience involves nothing less than our whole embodied presence, so the semantics of a descriptive phenomenology should automatically be considered a poor synecdoche for the actual 'being' or 'having been'. According to Thompson,

> Making aspects of experience explicit in this way unavoidably involves interpretation and the creation of meaning. … It also presupposes a pregiven background of tacit and unreflective experience that can never be made fully explicit. In these respects, accounts of prereflective experience are interpretive and not merely descriptive. (2007: 316–7)

Acknowledging that such tensions are inescapable, we might at least address and negotiate them by applying an enactive cognitive framework. Enactivism combines core interests of both phenomenology and cognitive studies, insisting that we recognize physiological uniqueness and a lifetime of singular experiences as being integral to how cognition is done. Enactivism 'takes seriously, then, the philosophical critique of the idea that the mind is a mirror of nature' (Varela, Rosch and Thompson 1991: 19) while suggesting that our consciousness is itself 'animated by precognitive habits and sensibilities of the lived body' (Thompson 2007: 24). Because it accounts for the psychological and biological realities of subjectivity by using data culled from cognitive, neurological and life sciences, enactivism accesses new and robust tools to disclose an increasingly precise phenomenology of experiences, including those which precede conscious awareness.

Noë, along with scholars such as Thompson and O'Regan, reframe skill as the very medium by which we come to know our world. We do not sense the world in any objective sense; rather, reality is disclosed in terms of our own abilities to act upon it, through a 'hands-on' practice of sense-making. For example, in the case of vision, the frequencies of radiation we sense as light are not inherently more 'perceptual' than other frequencies; similarly, the electrical impulses transferred by the optic nerve are not more 'visual' in content as compared to impulses from auditory, haptic or proprioceptive systems. However, because both light and light-receptive structures have materially consistent properties, they interact in reliably stable ways, which O'Regan and Noë refer to as *sensorimotor contingencies* (2001: 940). An individual thus hones a practical but implicit sense of how sensory change is likely to occur in response to intentional motoric actions (i.e. patterns of light striking the retina will predictably curve and distort when we shift our gaze away from it; patterns on the retina move in an expanding or contracting flow as we respectively move towards or away from the object).

So far as tactile sensation is concerned, when we touch a sponge, we feel it as being soft by implicitly recognizing patterns of sensorimotor interaction that become possible through our movement in relation to the object. According to O'Regan, Myin and Noë,

> the softness of the sponge is not communicated by any particular softness detectors in the fingertips, nor is it characterized by some intrinsic quality provided by the neural processes involved, but rather it derives from implicit, practical knowledge about how sensory input from the sponge currently might change as a function of manipulation with the fingers. (2005: 56)

The skills we exercise recede from the horizon of our experience as means by which other experiences are disclosed (Noë 2004: 20–2), but by exercising this implicit, skilled knowledge, one becomes aware of a contingently stable external world in relation to oneself. Therefore, even at the pre-symbolic foundations of our sensory experience, meaning arises through the practical mastery and skilful exploitation of these contingencies.

Such individuation does not obviate common processes of cognition across individuals. Sensorimotor contingencies may be reliably similar across most members of the species (e.g. the bio- and psycho-mechanics of my vision are much like those of the next near-sighted 30-something). However, if a framework for cognition is not given *a priori*, but emerges from the subject's lifelong mastery of her body's own sensorimotor contingencies in relation to her engagement with the outside world, then these processes are individually specialized. Physiological uniqueness and a lifetime of singular experiences are integral to how perceptual cognition is done. Perception lives 'no more in the sensory centres than in the motor centres' (Ansell-Pearson 2009: 156), but is the skilled negotiation of their complex relations by an embodied subject.

Because all our environmentally reactive sensory structures are concurrently disclosing interrelationships with the same world, we perceive the world and its objects as perceptually coherent and sensually rich, and we implicitly ally particular forms of perceptual practice with others. In the case of the sponge, tactile knowledge is disclosed in tandem with its visual percepts; consequently, both the tactile and visual percepts of the sponge are learnt as contingent properties of that object. The remarkable consistency of these experiences – their tight coupling – teaches us implicitly that a visually perceptible sponge will be soft. Accordingly, upon seeing the sponge, we anticipate the employment of those skills involved in tactile sensation, thus preparing and simulating neurological and muscular processes that we have implicitly learnt to run in such cases.[6] Consequently, we perceive objects in terms of our ability to act with them or upon them, highlighting the enactivist claim that 'traditional distinctions between action and perception arise only as the specialization of phases in an act of [skillful] sense-making' (De Jaegher and Di Paolo 2007: 489).

Training for disciplinary enculturation

Before applying this enactive account of sensorimotor perception to an analysis of training practices, we must bear two caveats in mind. The first is that while we may perceive objects as things to act upon, we perceive other subjects as agents whose actions we might emulate. Jeannerod notes that we often imitate others' actions without being conscious of an intention to do so, whether that imitation is obvious or happens covertly, at weak levels of neuromusculatory excitation (2006: 135). This autonomous behaviour helps us to learn by mirroring others' actions, and it also helps us to understand others' emotions (see Neal and Chartrand 2011) and action intentions (see Grezes et al. 2003): in perceiving and imitatively embodying familiar activity, we come to anticipate the intentions we have associated with our own embodiment of those actions. Patterns of learnt contingency may be increasingly complex, but our own psychophysical 'expertise' provides a general basis for recognizing unseen intentions underlying the perceptible facets of another's psychophysical experience.

The second caveat is that, unlike most subject–object interactions, subject–subject interactions don't necessarily enter into wholly lawful relationships. According to De Jaegher and Di Paolo,

> We don't experience the other-in-interactions as totally obscure and inaccessible, nor as fully transparent (like an object fully constituted by my sense-making activity), but as something else: a protean pattern with knowable and unknowable surfaces and angles of familiarity that shapeshift as the interaction unfolds. (2007: 504)

What we *do* come to know of others is an educated and embodied 'best guess', subject to further mediation in light of the others' actions, but provisionally stable enough to afford a practical knowledge of that other as an autopoietic cognitive system.

Social and cultural identities may likewise be known as an implicit system of practice, superficially established in terms of what is performed, but, as we saw in the earlier study of the expert dancer, more subtly and pervasively enacted in how we perform it. Movement studies by Gill and Kawamori, for instance, have found that subjects' gestures may be individually idiosyncratic, but tend to be culturally characteristic (Gill 2012: 114), while vision studies by Lewis and colleagues have found robust ethnic cultural bias in the minute perceptual strategies adopted by subjects looking at an image (i.e. attending more or less to the background context of visual scenes). Through diverse social, biological and experiential channels, cultures 'invade us and set our agendas' (Donald 2001: 298), influencing the dispositional enactment of our skilful sense-making practices while implicating uniquely salient domains of action.

Consequently, the acts, gestures or movement vocabularies that are embodied through the practice of a psychophysical discipline are significant both for the discrete intentions to which they refer, and for the way in which they disclose the deeper, more pervasive internal logics underlying the behaviour of individuals or diffuse social networks to which they belong. The deliberate coordination of embodied social practice may certainly yield highly technical, specialized forms of embodiment, but the non-verbal dimensions of a domain-specific institutional micro-culture will necessarily afford those skills in the context of its own disciplinary *telos*. According to Lizardo,

> The capacity to understand and 'grasp' the meaning and *telos* of action by other agents at an implicit, bodily level, without recourse to an explicit 'theory of mind' of other agents, coupled with the capacity to 'mirror' the action of others and engage in implicit imitation of the bodily techniques of others, provides a completely different perspective of what it means to be 'socialized' into the 'culture' of a given collectivity. (2009: 718, italics in original)

For example, in Wacquant's immersive study of boxing culture, the boxer in his gym is drawn into an enactment of the exercise he has learnt to perform, but also into a rhythmic synchrony with the other boxers in the space, seen and heard at all times through mirrors and in close proximity. Lizardo believes this exemplifies how training 'structures learning and processes of embodied simulation', amassing bodies within designated times and spaces 'to produce the collective synchronization of embodied rhythms that aid in the encoding and retrieval of practical information' (2009: 720). The practice of training provides of a 'complete network of relations [that] constitute the space of (physical, auditory and visual) exchanges' (Ibid: 719), implicitly and empathically embodied. Disciplined psychophysical training thereby inculcates in its subject a characteristic style of sense-making as much as it imparts any discrete skills.

Skill and insight in flow

While we may not wish to claim that all skilful performance is qualitatively 'fluid', Mihaly Csikzentmihalyi's theory of flow provides a valuable touchstone for describing an expert performer's experience of fluidic immersion in the task at hand. Csikzentmihalyi describes flow as an 'almost automatic, effortless, yet highly focused state of consciousness' (1997: 110), at once pleasurable and creatively productive. It takes place when our skills are being optimally challenged by a given task (Ibid: 111–12), and its phenomenological features include an intense or heightened concentration and attention on the activity at hand; a sense of distorted time; transcendence of the self; a reduction or loss of reflective self-consciousness; reduced or absent worry over failure; focus on the present moment; resilience against distraction; autotelic engagement in the activity itself; merging of action and awareness; and the feeling of being at one with the environment or activity. (Vervaeke and Herrera-Bennett 2013)

Flow phenomena may be observed in any number of activities, from rock climbing to bobsledding to improvisational jazz. For the actor, flow offers a canny description of what it means to 'be in the moment'.

Perhaps most helpfully, flow theory provides a mechanism for understanding how training supports novel expressions of skill by recognizing that the experience of flow is akin to an experience of creative insight. In straightforward instances of problem *solving*, knowledge and skill are selectively applied in light of familiar patterns in the information at hand (Bassok and Novick 2012), but *insight* is a form of problem *finding*: we differentiate information at hand within an otherwise ill-defined field of data, or see the problem in a new way (Steenburgh et al. 2012), which in turn, allows us to integrate that problem in terms of our capacity to act upon it. By reframing the problem, we also disclose the solution, which rushes to us as that Archimedean 'Eureka' or 'Aha' experience that is the hallmark of insight. A classic example would be in completing the nine-dot problem, wherein a subject is presented with a square grid of nine dots and asked to connect them using four straight lines, never lifting her pencil from the paper (see Steenburgh et al. 2012); most people struggle with the problem until they recognize that they can (and must) extend their lines beyond the imagined confines of the grid itself, quite literally 'thinking outside the box'. Insight is also helpful when one is engaged in a challenging psychophysical task like playing hockey, sparring with an opponent or improvising in rehearsal. The performer's immediate goals may be ill-defined and seldom articulated, but she arrives at an immediate understanding of the problem to be solved at the same time that she begins to act upon the solution. The performer's actions subsequently impact her field of play, so that each insight affords new knowledge and continually reconfigures the information sets she may perceive.

Csikzentmihalyi admittedly stumbles in modelling this process by imagining insight to be a natural by-product of the subject's intuition (1997: 102) while simultaneously insisting that our explicit knowledge, including 'the knowledge of the domain [and] the concerns of the field' is what sets patterns of unselfconscious thought (Ibid). This retains a misplaced emphasis on the hierarchical coherence of domain-specific skill sets (see also Sawyer 2006: 125), chafes against many of the concerns discussed above and mysteriously exempts practices of non-linear cognition from processes of internalization.[7]

Fortunately, John Vervaeke and Arianne Herrera-Bennett propose a refined model of creativity in flow that places intuitive insight at the very heart of the experience. They suggest that in one's experience of flow, each insight propels us towards perceiving new problems and insightful solutions in a positive feedback loop. Vervaeke and Herrera-Bennett refer to this feedback loop as an *insight cascade* (2013). This cascade is evident in performance as 'each new gesture' and each new insight 'thus apprehended-comprehended becomes in turn the support, the materials, the tool that makes possible the discovery and thence the assimilation of the next' (Wacquant, quoted in Lizardo 2009: 720). The discipline of one's skill provides a constraint for the subject's operations within a self-organizing system of practice, facilitating emergent flow behaviour while favouring challenging modes of skilled practice and coordinated forms of implicit learning. Understanding flow as insight cascade functionally links our account of creative expertise in unselfconscious performance with the implicit integration of psychophysical knowledge that takes place in training.

Cultivating flow through mindful practice

The application of this refined theory of flow to improvisational or theatrically devised work is comparatively direct. The term 'devising' does not so much encapsulate a tradition as act as a signpost for a broader nexus of practices where, as Turner and Behrndt would put it, 'the content, form and structure are determined as the process unfolds. The performance text is, to put it simply, "written" not *before*, but *as a consequence* of the process' (2008: 170, italics in original). Multiple creative strategies may be developed, applied or rejected throughout a single process, and devising's capricious multidisciplinarity affords idiosyncratic modes of participation through which spontaneity may be expressed and insight may be gleaned. Barton suggests that the devising performer requires a 'heightened sensitivity to possibility and a rigorous ability to exploit its gifts' (2005: 105), which marks her as a clear beneficiary of training that promotes creative flow.

Even in the absence of an overtly generative processes – that is, in performance for scripted texts where dialogue or movement is proscribed – skilful performance is rightly understood as 'a specific *how*' rather than 'a certain *what*' (Vervaeke and Herrera-Bennett 2013). Barba's early work with the Odin Teatret (a theatre troupe based in Denmark), for example, emphasized gymnastic training. Odin may have produced fine gymnasts, and its vocabulary of action served its intensely athletic production work, but it adopted these practices because its members believed the qualities attendant to mastery within a gymnastic sphere of engagement were useful beyond itself (*Physical Training* 1972). The actor develops an extra-daily habitus that may be employed in her inhabitation of a theatrical milieu, applying the enhanced perceptual acuity, physical prowess, flexibility and responsivity to a broader range of expression. She has fostered a cognitive style that proves adaptively beneficial to ulterior contexts.

Yet while we might practise diverse forms of psychophysical discipline to cultivate a skilfully unselfconscious mode of engagement for performance, the way in which we train may be optimized by incorporating a practice of mindfulness. Mindfulness is a domain-crossing cognitive style, a widely shared capability (Dane 2011: 998) whereby our consciousness 'operates *upon* rather than *within* thought, feeling, and other contents of consciousness' (Levesque and Brown 2007: 285, italics in original). We may be thought mindful when we are actively attending to the task at hand as we are doing it, de-automating instinctive performance in order to 'prolong that initial contact with the world' (Brown, Ryan and Creswell 2007: 212). Mindfulness may therefore be understood as 'a state of consciousness in which attention is focused on present-moment phenomena occurring both externally and internally' (Dane 2011: 1000).

If tasks are performed mindfully in training, one's awareness optimally touches on a full-body awareness of co-relevant contingencies between the subject's actions and the world they disclose. Both mindless performance of and fixed hyper-vigilance upon a task can severely limit the number of phenomena that one perceives externally, even if the missed phenomena would be relevant to the task at hand (Langer 1997: 44), but individuals practising mindfulness recognize more stimuli in their environment, even when those stimuli are especially subtle (Dane 2011: 1001). The same goes for internal sensations: mindful subjects demonstrate enhanced cognitive flexibility (Horan 2009: 211) and can successfully attend to emotions and physical sensations without coldly removing themselves from the experience (Brown, Ryan and Creswell 2007: 214; Langer and Moldoveanu 2000: 2). The shift for Type 2 analytic processing is subtle yet critical, as the analytic mind attends to the holistic breadth of activities being run through Type 1 processing, drawing attention towards what it is to inhabit a given task rather than towards any singular element of completing the task itself. Learning emerges from a holistic awareness of the moving body rather than from the micro-correction of discrete units. According to Zarrilli, 'Daily repetition allows the actor time to explore ever-subtler dimensions of the body, the mind, and their relationship in action' (2008: 29).

One's capacity to be mindful must also be trained in order to be skilfully applied, either as an ancillary benefit of work in other disciplines or in its own right in the practice of meditation, but it ultimately affords the practitioner greater control over the terms of their phenomenological experience, even insofar as they wish to cultivate flow. According to Horan, the intention to 'transcend informational boundaries' (a.k.a., functional differentiation) and 'integrate the transcendent experience, valuably, within empirical reality' (a.k.a., functional integration) is common to both meditative and creative insight (2009: 199).[8] Meanwhile, the cognitive task of meditation provides precisely the sort of open-ended, feedback-rich insight problems that emerge from more externally oriented processes of flow. This provides a template for the actor's ability to engage in creative flow while inhabiting a given role, such that proscribed actions form not a rigid scaffold but a rich and evolving milieu for the actor's creative sense-making.

Phillip Zarrilli's pedagogy is exemplary in its regard for mindful practice.[9] He begins training actors with a technically thorough and physically robust introduction to the practice of martial arts, particularly *kalarippayattu* and *taiqiquan*. Throughout, he encourages pupils to be mindful of the form, of one's breath, of the floor, of the surrounding environment and of the lines of flight linking where one's attention has been and where it is going. Though the forms rarely (if ever) enter exercises on acting work, he often suggests during acting exercises that students recall and once again embody that same state of immersive awareness that the form training affords.

Zarrilli aligns this subtle, holistic sense of affordance with the terms *qi* (from Chinese) and *prana* (from Sanskrit), as both an energetic principle and as 'a material actuality that can be awakened and raised through specific types of training ... [then] modulated and shaped for use within specific practices' (2008 19). It 'enlivens and quickens one's awareness, heightens one's sensory acuity and perception, and thereby animates and activates the entire bodymind' (Ibid). Vervaeke draws this link as well, suggesting that without negating the cultural and philosophical specificity of the concept '*qi*', we may nonetheless look to the phenomenon it circumscribes for a canny description of cognitive states linked into flow phenomenon (2011). Similar concepts may be traced in the work of Stanislavski, Chekhov, Grotowski and particularly Barba in his theorization of the *pre-expressivity*, which is reportedly cultivated in heavily codified styles of psychophysical training, and conceived as 'a pragmatic category, a praxis, the aim of which, during the process, is to strengthen the performer's scenic *bios*' (Barba and Savarese 1991: 188). This clearly alludes to a cultivation of *qi*-style affordances, describing pre-expressivity as an energetic liveness that underlies expressive performance. Consequently, where the acting score is not set, the trained actor may engage with both her skilled experience and her environment as they implicate an expanding field of expressive possibilities in real time, thereby enacting a sustained pattern of creative discoveries; where that the acting score is set, an attitude of affordance consonant with *qi* phenomena allows the actor to enter into a dynamic relationship with the dramaturgy, made present to her in a perceptible structure of words, actions and environments with which she might engage.

To be sure, training and enculturation neither inherently nor always lead towards sophisticated automaticity in the deployment of skill. The sort of fluid practice afforded by an embodiment of disciplinary *telos* is often experienced in fragments, perhaps as a raft of Type 1 processes supporting an overarching action-intention that remains under Type 2 control, or in fleeting moments of 'Von Kleist-ian' grace. However, with an enactivist model of skill and in the context of Vervaeke and Herrerra-Bennett's refinements, we can bring the straightforward performance of skills into conversation with the unselfconscious mastery and ancillary benefits it affords. Such a study demands we remain sensitive to the experience, intentions and actionable milieu of the individual subject, but identifies a common cognitive ground for behaviourally diverse, phenomenologically similar creative processes. Moreover, the term *qi* may account for that multi-adaptive strategy of mindfulness that supports a creatively masterful practice of skill, demarcating an openness to insight that is at once profoundly wise and nakedly responsive. If 'paradise is locked and the cherubim behind us', we thereby depend upon the gifts of training to 'journey around the world and see if it is perhaps open again somewhere at the back' (Von Kleist [1810] 1991: 238).

A Response

The Body in Mind

Catherine J. Stevens[10]

The three chapters written by Jackman, Utterback and Warburton encapsulate the excitement and necessity of an embodied cognitive science. Jackman analyses self-consciousness in performance and the liberty and artistry achieved when extensive, effortful training enables effortless automaticity and flow. The process is dynamic with the 'performer's actions subsequently impact[ing] her field of play'. Three strategies in actor training, built on bodily thinking and the actor as athlete, are outlined by Utterback – power posing, mental imagery and positive self-talk. Concepts and approaches from clinical psychology in general, and sports psychology in particular, have pertinence and relevance for acting and performance. Warburton argues for thinking in dance as dance enaction, illustrates the practical and communicative elements of making in dance and emphasizes cognition in dance not as capacity but meaning-making.

Each chapter touches on different aspects of human performance. The variety and complexity of the performance settings demand of cognitive science an interdisciplinary, multi- and inter-modal approach. For example, Warburton argues for a description of content – what is felt and thought – in dance, and proposes 'marking' as an example of a content-rich concept. Such content is necessarily multimodal. Strategies discussed by Utterback build student actor esteem and confidence by targeting particular postures, inner speech and imagery. Jackman notes that perception by an embodied subject entails skilful negotiation or the interrelationship of sensory and motor processes. The challenge presented by such subject matter is reminiscent of a personal quandary.

In 1999, I began a research project called 'Unspoken Knowledges' with dance writers Shirley McKechnie and Robin Grove in collaboration with, and partly funded by, professional dance organizations in Australia. Having completed a PhD on auditory cognition and memory for music, I was able to watch rehearsals in which a choreographer of contemporary dance and eight dancers collaborated to create a new dance work. The methods of auditory psychophysics that I had adopted for my laboratory and experimental investigation of music cognition fell short of what was needed to capture, let alone explain, the improvisation, problem finding and solving, decision-making, watching, showing, talking, moving, experimenting, discovering, creating, laughing, playing, generating, drawing, imagining, exploring, recording, recalling, forgetting, reproducing and describing, that happened each day in the dance studio. The processes I observed of making, refining, remembering and performing contemporary dance were at once visual, motoric, auditory, spatial, expressive, communicative, temporal and somatic. There was no doubt that the phenomenon was cognitive and intellectual, driven by and exuding meaning. It was also highly technical and physical. The behaviour and artistry I observed in the studio, like the scenarios described so clearly by Utterback, Jackman and Warburton, epitomized embodied cognition. My observations led to me to understand that perception and action are coupled. The body moves intelligently. Intelligence is bodily. Actions constrain or shape thought. Thought guides action. The processes of perception–action–cognition are cumulative and dynamical.

My colleagues and I began by simply describing the creative process (Grove, Stevens and McKechnie 2005; Stevens et al. 2003), learning from the ethologists Konrad Lorenz and Niko Tinbergen the importance of observation. From observation came hypotheses that could be put to eventual empirical test.

A second project with McKechnie and Grove, 'Conceiving Connections', investigated audience reactions to some of the contemporary dance works that had been created as part of the first 'Unspoken Knowledges' project. We learnt first hand of the embodied and sometimes implicit nature of knowledge in dance. It became apparent that words are not always the best way to communicate or elicit dancerly knowledge (Glass 2006). Accordingly, we started to use more indirect indicators of perceptual and cognitive processes in our experiments, including continuous ratings of emotional response or engagement (Stevens et al. 2014; Stevens et al. 2009), eye movements as proxies for visual attention (Stevens et al. 2010), and paradigms based on the implicit (unconscious) learning of an artificial grammar (Opacic, Stevens and Tillmann 2009). Recently, we have used dance-like material to probe short-term memory for form and motion (Vicary et al. 2014).

A series of theatre- and laboratory-based experiments with my colleagues showed that it was again observation that gave rise to insights into what Warburton would call 'meaning-making'. I had an opportunity to conduct an experiment on memory for dance material that had been learnt in a span of three to thirty years ago. Four mature dancers came together to document

'Dance-Drama' exercises that had been created in Australia from the 1950s to the 1980s by Margaret Barr (Stevens, Ginsborg and Lester 2011). What was striking in the dancers' recall of the material was the diversity of ideas recalled – the politics and social issues of the time, memories of the spatial layout, location and smell of the dance studio, the timbre and prosody of the choreographer's voice, Barr's pose and presence, and, through mental time travel, the dancers' recall of their own sense of self at that earlier time. The dancers' self-reported recollections resonate with Utterback's concepts of physicality and command from posture and imagery in many modes.

Barr's dancers had been asked to recall the dance-drama exercises alone and without music but more movement material was reproduced when the dancers recalled collectively and when the accompanying music was heard. It seemed that even more movement was retrieved from memory when the dancers reproduced exercises that they said they 'loved doing'. Observing these four dancers showed that memory for dance is distributed rather than residing intact in individual brains and bodies. By its association with all of the sensory systems, with verbal memory and with emotion, dance memory is full of meaning.

Warburton's account of dance marking brings to light many of the key elements that characterize dance performance and memory – body posture, timing, balance, chunks of choreographic phrasing and subtle features of performance expression. His empirical study provides support for marking as a way to reduce cognitive load. Competition for resources – those needed for expert performance of complex material and those for thinking more abstractly about material to be memorized – is managed, he says, by marking. It bestows cognitive benefits, not only through economy of load but by facilitating communication and cohesion during the rehearsal process.

In his chapter, Jackman notes the disjunction between practice systems and performance; similarly, Utterback comments on significantly more time being spent training than performing. These observations bring to mind scales, arpeggios or technical work for the musician which are essential but rarely part of performance. Exceptions are 'etudes' and 'studies' valorizing technique in and of itself. Practical systems like musical scales, ballet barre exercises or actors' elocution and tongue twisters build technique and good habits. Technique in the arts is honed and underpinned by acquired expectations, anticipation and spatial and other kinds of awareness. Improvisation, of course, often aims to break habits. Jackman refers to the benefits of mindfulness and *qi* for the performer. The science is emerging too, that provides some evidence of their impact on electronencephalography (EEG) recordings of brain activity (e.g. Berkovich-Ohana, Glicksohn and Goldstein 2014; Cahn and Polich 2006).

Dance has been seemingly part of human cultures since antiquity, and its ubiquity implies that there is some evolutionary advantage conferred by dance. While some seek behavioural evidence (e.g. Bachner-Melman et al. 2005; Brown et al. 2005), we can theorize that 'ontogeny' of a particular work of art 'recapitulates phylogeny' with movement as an evolutionarily early form of expression. In the acting, music or dance studio, germs of ideas evolve into full-blown works of art. Ideas are expressed in many different modes and gestural forms but it is *movement* – of limbs, air pressure, bodies, vocal tracts – that is core. Much later, ideas may be codified, notated, and even mass distributed. But the initial generative improvisation and creative process is often non-verbal and silent, collaborative, distributed, dynamical, self-organizing and potentially adaptive (McKechnie 2005; Stevens and McKechnie 2005b).

At the level of the individual, developmental psychologist Esther Thelen (1995) proposed that an infant's earliest cognition, its coming to know the world, is through movement. Stevens and McKechhnie (2005a) speculated on a 'critical period' for dance-like movement with an apparent loss of freedom in movement for some perhaps about the time when the complex task of learning to read consumes much of a child's cognitive resources (Burnham 2003). Is there a critical period for dance? Akin to acquiring a second language as an adult, might it be the case that the 'vocabulary' of dance is learnable but perfecting the 'grammar' of movement is elusive? Vocabulary may be acquired explicitly and consciously whereas proficiency with grammar may require extended 'mere exposure' for implicit or unconscious learning to take place (Opacic, Stevens and Tillmann 2009; Ullman 2004). An alternative hypothesis is prompted by the concept of self-consciousness in acting (Jackman). If it is the case that the sense of self takes years to develop (e.g. Fivush 1994; Wang 2001), then a critical period for dance may terminate with the child's increasingly developed sense of self and, relatedly, instances of self-consciousness.

As a kind of oral tradition, and before the development of written language, dance and music were means of communicating myths and identity between groups and generations. Rhythm in both movement and music would have served as a mnemonic in such settings. A focus on the human capacity for rhythm and timing, often at its most exquisite in music, dance, and sport, is driving a current surge of research in cognitive neuroscience. Using methods that measure behavioural responses (accuracy, reaction time) and patterns of neural stimulation, cognitive neuroscientists are investigating links between synchronizing to a regular beat or rhythm, social factors, and pair or group bonding. For example, Kirschner and Tomasello (2009) hypothesized that synchronizing body movement to an external beat, as seen in dance and music, evolved in human cultures as social activities. In an experiment with children drumming spontaneously, they showed that young children spontaneously synchronize to a beat at an earlier age when those children are in a social context. That is, when they are in the presence of a human partner producing the rhythm rather than with a drum machine or a loudspeaker.

The *advantage* of synchronizing with another has also been demonstrated experimentally. Acts of joint music making and movement in four-year-olds increased the children's subsequent spontaneous cooperative and helping behaviour (Kirschner and Tomasello 2010). At the neural level, the human beat-finding capacity is underpinned by entrainment of neurons at multiple levels in the brain (Merchant et al. 2015). Phillips-Silver and Keller (2012) theorize about different kinds of entrainment associated with synchronization distinguishing time-based or temporal entrainment from affective entrainment. Temporal and affective entrainment, they argue, are significant for human behaviour and human interactions across the lifespan. Such entrainment manifests in infant–caregiver interactions and turn-taking (e.g. Dissanayake 2000; Malloch and Trevarthen 2009) and is also likely to operate within groups of actors, singers and dancers and between performer and audience (e.g. Leman et al. 2009; Moran 2014). In the context of contemporary dance, we have begun to examine effects of the social on improvising and have compared dancers improvising alone with improvising in familiar or unfamiliar dyads (Stevens and Leach 2015).

The three chapters in this section are not only deep with theory but also have tangible applications to the performing arts, public speaking, coaching, training, teaching and sport. Jackman's analysis of flow applies to acting in particular and performance more broadly. The goal is for 'a skilfully unselfconscious mode of engagement for performance'. Suggestions are made where the acting score is not set and where it is set. In the former, the actor's skilled experience and environment inspire an expanding field of expressive possibilities; in the latter, the actor enters into a dynamic relationship with the dramaturgy affording words, actions and environments. Utterback presents a strong case for strategies to prepare student actors for the physical and emotional demands of performing under the pressure of observation. Methods of power posing, imagery and self-talk focus attention and energy, bolster self-confidence and, over the longer term, bootstrap habits for maintaining confidence and esteem.

Contexts and issues presented by Utterback, Warburton and Jackman assume further development of interdisciplinary inquiry. There is potential for new emergent approaches that arise from the complementarity of qualitative and quantitative methods and the nexus of arts, science and the humanities. Studies with culturally diverse groups, materials and settings, as Warburton's research in Asia attests, will further the growth of theory and practice. Almost a decade ago, Noice and Noice (2006) set out the importance of meaning and meaning-making in healthy cognition and ageing. These three chapters have potential for immediate impact on theory, on performance training and practice. Conceivably, acting, dancing and musicking opportunities, where we improvise, forge new neural connections and make meaning throughout our lives could have incalculable long-term benefits in delaying, even preventing, cognitive decline in our global ageing population.

Part Three

Situated Cognition and Dynamic Systems: Cognitive Ecologies

If embodied cognition locates thinking in the whole organism, as inseparable from the actions of that organism, embedded cognition argues that cognition uses the environment. The prop table organizes what is needed by the actor so that between scenes the actor does not need to look at the table, remember what he needs for the next scene and search for it. The stage manager marks the paper on the table so that the prop goes in the same place every night – often tracing the edge so the hammer goes where the hammer silhouette is. This allows us to 'travel information light', using prompts in our environment to guide our actions in the moment. We can take advantage of this – as the stage manager does – and alter our environment to support the offloading of cognitive tasks, which David Kirsh and Paul Maglio call 'epistemic action' as opposed to 'pragmatic actions' which make an environmental change because it is desirable in and of itself, not because of what it affords. We create tight feedback loops between our actions and the world they act on and with, in order to operate quickly and efficiently.

Situated cognition sees these 'coupled systems' as cognitive in their own right, taking the mind beyond the body. In this view, thinking is what happens between agents and their environmental tools, coupling seemingly discrete units in a system that extends what and where we imagine the 'mind'. If our environment becomes part of the cognitive act, I am only as smart as the system that surrounds me.

For decades, scientists studying ants attempted to locate in the ant an intelligence that modelled our own; how we thought of our thinking – with language areas, representation and navigational maps – was how we tested their thinking. This assumption limited what we could learn about our thinking and completely missed the intelligence of the ant.

Ants use a variety of cues to navigate, such as sun position, polarized light patterns, visual panoramas, gradient of odors, wind direction, slope, ground texture, step-counting ... and more. Indeed, the list of cues ants can utilise [sic] for navigation is probably greater than for humans. Counter-intuitively, years of bottom-up research [assuming simplicity and studying the trait, rather than testing for a top-down explanation for the presumed sum-total] has revealed that ants do not integrate all this information into a unified representation of the world, a so-called cognitive map. Instead they possess different and distinct modules dedicated to different navigational tasks. These combine to allow navigation. (Wystrach 2013)

In other words, ants glean extraordinary information from their environment and they act on it but they do not 'think' about it.

This paradigm shift to a situated view of cognition is of such magnitude, requiring new language and a plethora of new terms, that it can be overwhelming. 'Situated cognition' has been used to encompass terms such as 'embodiment, enactivism, distributed cognition, and the extended mind' (Robbins and Aydede 2009: 3), all attempts to understand and articulate the complex, dynamic nature of cognition. All commit to the embodiment of cognition. All perceive some degree to which cognition exploits the 'structure in the natural and social environments' (Robbins and Aydede 2009: 3). All argue for rethinking the way cognition transcends the boundaries assumed to exist between organisms. There are also differences between some of these views, with some of them making much stronger commitments and claims than others. For example, Edwin Hutchins explains that extended mind

> picks out a kind of cognition. In the extended mind view, mind may sometimes extend beyond the brain, and sometimes it does not. ... The extension of the mind is manifest in links and relation that cross the usual boundary of the mind container. Second, extended mind assumes a center in the cognitive system: the organism (or the organism's brain), which is the normal mind container with respect to which cognition can be said to extend. (2014: 36)

Distributed cognition, however, is a way of thinking about all cognition:

> Distributed cognition begins with the assumption that all instances of cognition can be seen as emerging from distributed processes. For any process there is always a way to see it as distributed. In practice this implies that wherever we find cognition, it will be possible to investigate how a process we call cognitive emerges from the interactions among elements in some system. (2014: 36)

The value for theatre and performance scholars is not in which of these theories ultimately persuades the most scholars or scientists, the value is in how we might rethink our categories of research.

As Evelyn Tribble (2011) points out, a Romantic notion of the Lone Genius led to centuries of interest in Shakespeare: who he was, how he did what he did and what he wrote or didn't write. Applying research in situated cognition, Tribble suggested that perhaps there are signs of cognition in the system and network within which Shakespeare's theatre happened. Imagining Shakespeare's Globe as an environmental tool used by the players to facilitate the heavy cognitive task of performing five to six different plays per week, Tribble argues that the players and environment of the Globe can be thought of as a system that creates and perpetuates cognition. Tribble looks for an answer *not* in the individuals, but in the system. Writing with cognitive scientist John Sutton, Tribble argues for thinking about cognitive ecologies: 'Cognitive ecologies are the multidimensional contexts in which we remember, feel, think, sense, communicate, imagine, and act, often collaboratively, on the fly, and in rich ongoing interaction with our environments' (2011: 94). A focus on cognitive ecologies stops extracting individual 'thinking' agents as figures from the 'ground' within which they stand and work; cognitive ecologies examines the situated and distributed system of cognition.

The chapters in this section explore variations on the implications on this different way of looking. Tribble shows us a number of different performance environments that 'enskill' performers, from the orderly kitchen of a top chef to the different theatrical structures that controlled the cognition of Shakespeare's Globe versus the actor-manager system that came to be after the Restoration. She places ideas of mastery and expertise within a situated cognitive environment. Sarah E. McCarroll takes up the idea of body image and body schema from Shaun Gallagher and others to argue that ideas about how the body moves are culturally and historically embedded through clothes. She looks at how theatrical performances in nineteenth-century England taught the middle and upper classes how and where bodies could and should be. McCarroll's historical reading suggests that such performances are an opportunity to look at the situated body, recognizing that bodies carry the marks of their clothes, culture and the requirements of deportment. Matt Hayler finds in new forms of dance evidence of new ways of looking afforded by technology. From a post-phenomenological perspective that insists that humans are 'mediated and co-constituted by our engagement with material artefacts' (235), Hayler sees the bodies of certain dancers thinking with and through technology that has provided them with new ways of imagining the body and what it can do. He posits a visual grammar that controls the sense we make out of what we see and that can be expanded or shifted, depending on what and how we see. As we change our environment, we change our thinking.

The ecologies of performance – whether it is London's Victorian theatre or the dancing, repeating, watching of an internet generation – create cognitive systems. Seeing our theatre and performance history as always a part of an ecology enables us to ask and answer new questions about what happens when *these* bodies take *this* stage *here* and *now*.

Distributed Cognition, Mindful Bodies and the Arts of Acting

Evelyn B. Tribble

One of my guilty pleasures is the reality cooking show, especially 'MasterChef Australia'. This programme features a group of amateur cooks who are faced with a range of tasks, from cooking their favourite comfort food to experimenting with molecular gastronomy. The programme's stated goal is to crown the best amateur cook in the country. The contestants are extremely talented, and the finalists are extraordinarily adept at completing difficult, time-pressured and unfamiliar cooking tasks. They have high levels of cooking knowledge and in most cases have built skills over years of home cooking.

That said, even when they compete on an individual basis (rather than in the dreaded 'team challenges'), they are not cooking alone. Rather, they work in highly managed environments that have been carefully and intelligently designed. They work at rows of benches, surrounded by clean, high-tech equipment, sharp knives and plenty of cooking space. When they compete in particularly difficult challenges, they are provided with pre-measured ingredients. They can see what other contestants are doing, and the judges and guest chefs patrol the space, dropping hints and answering questions. Each competes on his/her own, and some of course do much better than others, but all of them are supported by the taken-for-granted scaffolds that the show provides.

The relationship between the chef and his/her environment becomes clearer when the contestants are suddenly parachuted into another space. Sometimes they are embedded in an existing working kitchen, where they often flail about until they are more or less integrated into the existing structure of the station (often with the overt or covert intervention of the professional kitchen staff). Usually instructed by a veteran chef, they begin to cope with the repetitive time-pressured tasks of 'service'. Even more difficult is the taking over an entire kitchen; in a recent show, the top five contestants were tasked with working in a small pub kitchen. Accustomed to the spacious, tidy and individually oriented benches on the set, they were almost completely unable to cope with the tiny space and limited equipment. They constantly jostled one another, vied for the same equipment at the same time and generally produced complete chaos. In particular, the task of effective coordination was beyond them. Accustomed to simply finishing their own dishes within a prescribed time frame, they were flummoxed by the need for precise timing, repeatedly underestimating cooking times so that some dishes became cold while others were barely finished. Only the constant haranguing of superstar chef Heston Blumenthal finally restored some order to the process. Even so, contestants who performed brilliantly in the well-appointed and managed test kitchen floundered when confronting the unfamiliar design and demands of the professional space, especially because they were not inducted into such a space as novices, as would be the normal practice, but were expected to perform immediately at a higher-order level.

Distributed?

I begin with this extended description as a way of thinking about the relevance of Distributed Cognition for conceptualizing performance, particularly the way that expertise can be seen as a 'culturally extended' phenomenon (Menary and Kirchhoff 2014: 610). Both aspiring chefs and aspiring actors develop expertise within carefully engineered environments that foreground some tasks and place others behind the scenes, so to speak. As I have written elsewhere (Tribble 2005), early modern players coped with seemingly overwhelming cognitive loads by being embedded within smart structures – physical, social, material, including the regimes of training and education that undergirded their practice. Playing companies were in fact building such structures as they went along, in an instance of cognitive niche construction.[1] In niche construction, humans engage in what Kim Sterlney and others refer to as 'epistemic engineering': 'organizing our physical environment in ways that enhance our information processing capacities' (Sterelny 2012: xii).

  The model of Distributed Cognition allows us to see how a complex activity such as performance is spread, smeared or extended across mechanisms such as attention, perception and memory; the experience of training as it is sedimented in the body, social structures and the material environment. Edwin Hutchins argues that 'distributed cognition is not a kind of cognition; it is a perspective on all of cognition' (Hutchins 2014: 36). This is an important distinction to make. In one sense, it is obvious that a collaborative activity like theatre or sport is 'distributed' across agents with different areas of expertise and knowledge. But this is a relatively weak claim; as Hutchins notes, the point is that thinking itself is a distributed process, although the 'boundaries of the unit of analysis for distributed cognition are not fixed in advance' (36).

Distributed Cognition also has affinities with the philosophical account of 'Extended Mind', best known from Andy Clark and David Chalmers's 1998 essay of the same name. Shaun Gallagher's account of the history of the term 'Extended Mind' rightly shows that it arose from an intervention into a relatively intra-cranial view of mind and that it still retains an emphasis upon material/technological extension (via, for example, iPhones and notebooks) rather than on the social world. Gallagher proposes a model of socially distributed cognition, arguing, 'If we think of the mind not as a repository of propositional attitudes and information, or in terms of internal belief-desire psychology, but as a dynamic process involved in solving problems and controlling behavior and action – in dialectical, transformative relations with the environment – then we extend our cognitive reach by engaging with tools, technologies, but also with institutions' (2013: 7).

Such a model lends itself to a variety of historical and theoretical approaches to performance. It predicts historical change and variation as new configurations of material and social practices emerge. A simple example might be the dramatic shifts in the theatrical enterprise in England following the period of the closing of the theatres (1642–60). Playbooks, parts, costumes and properties disappeared, of course. But even more importantly a system for cultural transmission of knowledge was lost. Before the Restoration, boys were apprenticed to companies in a more or less systematic fashion that ensured that knowledge and practices were passed on. Like midshipmen inserted into the smart environment of the naval vessels, novice actors benefited from intelligent structures: 'The task world is constructed in such a way that the socially and conversationally appropriate thing to do given the tools at hand is also the computationally correct thing to do. That is, one can be functioning well before one knows what one is doing, and one can discover what one is doing in the course of doing it' (Hutchins 1995: 224). The break in theatrical practice eliminated this system, and women began to play female roles. Perhaps at the outset, performing plays may have been a bit like being thrust into a professional restaurant with little training. As this new theatre evolved, however, new forms of practice – new conventions of the use of space and 'scenes' in the indoor theatres, the gradual establishment of actor-manager organizational structures, and revisions of the canon of drama to coincide with new conceptions of taste and decorum – emerged.

Cognition?

I suspect that words such as 'Extended', 'Distributed' or 'Embodied' might be more comfortable for theatre professionals than words such as 'Mind',

'Cognition' or possibly even 'Thought'. The latter words are often viewed with suspicion in discourse around acting training and methods. Actors often fear being 'in their heads', approaching acting intellectually rather than experientially. Indeed much of the discourse around acting is preoccupied with binaries such as inside-out ('method') versus outside-in ('technique'), thinking versus doing, emotion versus reason, and so on.[2] Rebecca Schneider, in *Theatre and History*, conjures up the image of the 'venerated teacher of naturalist acting who unwittingly may have fueled the anti-intellectualism of her student by saying, "You cannot think *and* do"' (2014: 33). She quotes the following passage from Sanford Meisner, whose Stanislavskian-inflected training methods had enormous impact on mid-twentieth-century American actors:

Student	I'm getting the feeling. Don't think – do!
Meisner	That's a very good feeling. That's an actor thinking. How does an actor think? He doesn't think – he does!
Student	Right.
Meisner	That's a good feeling. (Meisner and Longwell 1987: 57)

Meisner writes that his method is 'based on bringing the actor back to his emotional impulses and to acting that is firmly rooted in the instinctive. It is based on the fact that all good acting comes from the heart, as it were, and that there's no mentality to it' (37). In theories such as this, the body, the instincts and the emotions are arrayed against the mind, reason, 'mentality', the 'head' and 'cognition'. Meisner's point is probably meant to be as much a prompt or a heuristic as a literal proscription against intellect; nevertheless, the idea that the performer who eschews thought is someone more authentic than the performer who is 'in her head' is an enduring myth. A related phrase is 'thinking with the body'; this is often used in the performance arts. It is often deployed in a way that is intended to reverse traditional mind/body distinctions and to privilege a form of so-called bodily intelligence: 'In performance, the body is smart, intelligence puts on flesh – and we are smartest when we stop thinking' (Huston 1992: 6). But what precisely does it mean to 'stop thinking' in this context?

These are instances of a widely held belief that philosopher Barbara Montero has called 'the maxim': the assumption that thinking and doing are incompatible, and that particularly within a time-pressured performance, conscious attention to one's action inevitably degrades performance and spoils 'flow' (2007). This 'maxim' is particularly influential in two domains: sport and theatre (interestingly, it is much less common in chef lore, in which students are constantly enjoined to 'THINK!'). In both areas of activity a concern that over-thinking or over-intellectualizing will degrade performance and interrupt flow is common, as witnessed by Huston's injunction to 'stop thinking'. The great film and stage actor Paul Muni expressed the sentiment in this way: 'You know how to roller skate, don't you? Well, if someone asks you, how do you do it, what is your answer? You don't know. You just do it. You've learned how. You can be more adept by practising. But if you try to think about what you are doing while you are actually rolling along in the park, if you look down at the wheels and attempt to puzzle out what's going on, you fall down' (Eustis 2008: 19–20). In the domain of sport, but in a similar vein, the catcher Yogi Berra purportedly said, 'Think? How can you hit and think at the same time?' This observation is widely ascribed to him but dubiously sourced; nevertheless, the incompatibility of thinking and doing is a well-established truism in both sports and acting lore.[3]

These dualistic formulations fit nicely into the existing debate in the literature on expertise, which itself is structured around competing ways of explaining the phenomenon. The first of these is associated with K. Anders Ericsson, a psychologist at Florida State University, who has been studying the nature of expert performance for several decades. Ericsson's model stresses the 'expert's representation and memory for knowledge' (2007: 61). In this model, 'mere experience' is not sufficient; 'all the paths to expert performance appear to require substantial extended effortful practice' (61). Ericsson's work was popularized in Malcolm Gladwell's best-selling book *Outliers*, which put forwards the so-called 10,000-hour rule, the conclusion that even for the most talented individuals, 'ten years working and practicing' is necessary to become an expert (60). This model tends to stress declarative knowledge and conscious control. Ericsson and others in this tradition, including Roger Chaffin, who has written influential accounts of musical expertise, describe experts as constructing 'detailed, overarching cognitive frameworks constructed and consolidated over the course of many hours of practice' or a 'declarative roadmap' as the essential element in expert performance (Geeves et al. 2013: 67).

While Ericsson's work has been highly influential, it has been challenged as a 'cognitivist' model, excessively dependent upon conscious control. The best known of these challenges has been mounted by Hubert Dreyfus, working within the phenomenological tradition. Dreyfus argues that only relatively inexperienced performers rely upon cognitive roadmaps such as those described by Ericsson and Chaffin; instead, he claims that the true marker of skill is best described as 'mindless coping'. Repudiating the top-down approach of Ericsson and Chaffin, Dreyfus asks, 'Can philosophers successfully describe the conceptual upper floors of the edifice of knowledge while ignoring the embodied coping going on on the ground floor; in effect declaring that human experience is upper stories all the way down?' (2006:43).

That is, his model might be described as 'body up' rather than 'mind down', and is thus much more closely aligned with the 'outside/in' theory of acting. For Dreyfus, highly developed skill is less a result of cognitive rule-following than it is a form of 'absorbed coping' (48). Indeed, Dreyfus argues, as performance improves, the cognitive roadmap can be abandoned in favour of an embodied engagement with the world that emphasizes flow and the 'switching-off' of conscious monitoring (Dreyfus 2002).

As Andrew Geeves, John Sutton and others have suggested, such an account has intuitive appeal; the idea of flow resonates with experiences of fluid performance in such domains as sport, debate, acting and dance (Sutton et al. 2011). But this is only a partial account of 'thinking with the body'. The wholesale rejection of 'mindfulness' in skilled activity is 'built on over-reactions to ultra-cognitivist intellectualist or rationalist theories' (87). Simply reversing the dichotomy while being both 'culturally and intuitively appealing' (90), sells short the fundamental intelligence of skilled action: 'Certain patterns of behaviour [sic] which might appear stably chunked, automated, and thus inflexible are in skilled performance already and continually open to current contingency and mood, past meanings, and changing goals' (96).

Sutton and Christensen offer a middle way between cognitivism and mindless coping that they describe as 'Mesh'. In this model 'high order control plays a key role in virtually all action – not just action that involves explicit reasoning … expert awareness will be selective, highly shaped to task demands, and may often "roam" or "float" as it flexibly and anticipatively seeks out important information' (Christensen). So when, for example, a cricket player says that 'in batting you have to be mindless', he is referring not to the absence of thought, but to a certain deployment of attentional resources, 'employing sophisticated forms of attentional control and self-regulation' (Christensen). So too for the actor, who must banish certain forms of thought and harness others. In *Acting Power*, Robert Cohen refers to two kinds of thinking on the part of the actor: preparatory thinking as she readies herself for the role, and in-performance thinking, which, in an ideal situation, is 'aligned' with the actor's action (2013: 15–16). Cohen's quotation from Ralph Richardson nicely encapsulates the nature of the 'meshed' thinking of the expert actor:

> You're really driving four horses, as it were, first going through in great detail the exact movements which have been decided upon. You're also listening to the audience, as I say, keeping if you can very great control over them. You're also slightly creating the part, insofar as you're consciously refining the movements and perhaps inventing tiny other

experiments with new ones. At the same time you are really living, in one part of your mind, what is happening. Acting to some extent a controlled dream. ... Therefore three or four layers of consciousness are at work during the time an actor is giving a performance. (2013: 1)

One approach to studying how such meshed cognition occurs is the topic of 'marking' in dance. David Kirsh's Interactive Cognition Laboratory at the University of California, San Diego, has been exploring ways of using what Kirsh calls 'cognitive ethnography' to research such questions *in situ*, so to speak. Kirsh employs a mixed method that involves video observation and analysis and self-report on the question of how skilled dancers use mindful bodily practices to prepare and perform. Its acute empirical and theoretical examination of the often fuzzily defined concept of 'embodied cognition' and its collaborative cognitive ethnography is a powerful model for joint projects across disciplines.

Kirsh studies the practice of 'marking' in dance, a form of 'partial physical simulation' of a dance 'when marking, dancers use their body-in-motion to represent some aspect of the full-out phrase they are thinking about' (179). This might include executing a partial movement with the entire body, using a sweep of the arm or using the hand to 'stand-in' for the body. As one dancer says, marking for oneself (as opposed to marking for others to explain a movement) is 'a retrace, a scrape in your head' (181). In accordance with his overarching research platform of Distributed Cognition and skilled interactive environments, Kirsh argues,

> Through *externalization,* dancers are able to *attend* more effectively to difficult aspects of their movement than by mentally simulating that movement alone. Attention works differently on the outside than the inside. Physical structures have more details to cue off of and monitor than internal ones (199). ... People often think by means of an interactive strategy of creating external structure, then projecting meaning onto that structure, and then creating additional external structure. (207)

Such synecdochical movements are also performed by actors just before performance, as they run through their parts in compressed form, and by athletes who rehearse a partial movement, sedimenting a coaching cue into their bodies.[4] Kirsh's research is one example of ways that skilled performers think with the body, using both online and offline resources, and employing both somatic and environmental resources to think of movement.

Andy Clark has written that human beings extend or distribute cognition to make 'the world smart so that we can be dumb in peace' (Clark 1997: 180).

This provocation may make it seem that distributing or offloading thought is a form of mindlessness. But there is no inconsistency in seeing cognition as 'distributed' and in viewing expert performance as mindful practice. High-level, high-pressure, high-stakes performance is among the most cognitively demanding of all human activities and occurs within cunningly designed cultural frameworks and institutions. Agents within such a system do not behave mindlessly; rather, a distributed cognitive process allows focus to be thrown precisely where it is most effective. Studying culturally complex cognitive ecologies allows insight into the strategies used to structure and coordinate novel forms of skilled group action engaged in 'meshed' thinking.

The Historical Body Map: Cultural Pressures on Embodied Cognition

Sarah E. McCarroll

Clothes, proverbially, make the man, *do* something to the body, change its material reality. Beyond proverbs, however, clothes influence the ways we experience our bodies as structures with the capacity for movement and gesture. In this understanding, clothes literally *make* the man. Clothing has an impact on the way we live in our bodies, on our beliefs about how a body can and should move or gesture, on how our bodies live in and experience the world. The questions of how our bodies are structured, both mechanically and cognitively, and how they absorb and reflect the pressures of the society they inhabit have profound implications for our understanding of how embodiment is and has been integrated into the larger cultural space.

In *How the Body Shapes the Mind* (2005), Shaun Gallagher explores the ideas of body image and body schema, terms in use in psychology, philosophy, phenomenology and a range of other disciplines since the late nineteenth century. As Gallagher points out, the terms have been used with nebulous, contradictory or overlapping definitions, leading to conceptual muddiness in explorations of embodied consciousness (2005: 19–23). Here I use elements of Gallagher's clarified definitions of both body image and, to a larger extent, body schema to underpin my own proposition of what I term a historically situated and culturally pervasive body map. The features of this body map, I argue, are imposed on the landscape of the body via the structure of clothing and the dictates of social usages. I locate this theoretical exploration in a discussion of the communicative power of performers' bodies on the late-nineteenth-century theatrical stage. Through an examination of J. M. Barrie's *The Admirable Crichton*, especially the costume and carriage of Irene Vanbrugh's Lady Mary Loam in the 1902 original production, I explore ways in which the features of a particular body map were made manifest, and suggest ways in which the body map may be understood as the graph of a society's understanding of class and gender. Furthermore, I suggest that the theatrical environment creates a cultural space in which clothed actors' bodies participate in a transmission of social priorities.

Clothes make us. Not just for the eyes of others, but in subjective cognition, in our own mind's eye. Or rather, to follow Gallagher's rejection of the Cartesian mind/body duality, since the mind can only be structured from the reference point of the lived body, clothing makes the mind *by* making the body. In Gallagher's argument, our conscious and intentional experience of the world is unavoidably impacted by the inescapable fact that our perceptions are always embodied. If the body is understood as a kind of stage, on which performances of our cognitive structures occur, Gallagher (2005: 3, emphasis in original) extends the metaphor to ask, 'about the *structuring* of consciousness, and the role that embodiment plays in the structuring process. How does the fact of embodiment, the fact that consciousness is embodied, *affect*, and perhaps *effect*, intentional experience?' Gallagher (2005: 59) expands the impact of embodiment well beyond the limits of the human form: 'At stake here is not only how I experience my body, but how I experience the surrounding world,' he says. 'There is no disembodied perception.' Because perception of the world is embodied in the same way that perception of self inescapably is, there is an essential relation between movement and cognition; how we move through the world cannot help but impact the way we think about the world and our bodies in it. In fact, Gallagher and I both 'treat movement as a *constraint on* rather than a *cause of* perception' (2005: 8, emphasis in original). This embodied perception is constituted by two different mechanisms of bodily organization, the body image and the body schema.

For Gallagher, the body image is 'a system of perceptions, attitudes, and beliefs pertaining to one's own body' (2005: 24). Central to his definition of body image is the idea of intentionality, that is, the directed attention of the subject towards the body part in question. Within this 'complex set of intentional states and dispositions', is contained 'perceptual experience of [the] body; … conceptual understanding (including folk and/or scientific knowledge) of the body in general; and … the subject's emotional attitude toward [the] body' (Gallagher 2005: 25). The activation of the body image requires purposeful attention to the body part in question. Because of this required focus, the body image can never be fully holistic; while a subject has an image of the entire body, this image is not called into consciousness as such, but as the perception of or attitude(s) towards the singular body part that is in focus. From this definition of body image I borrow the idea that absorbed understandings of the body, drawn from folklore, scientific knowledge or socially pervasive truisms, have meaningful impacts on embodied experience. The body image is more malleable and obviously influenced by individual psychology and social pressures than the body schema, but as the performance of elements of the body image become pre-conscious, those elements may become part of the body schema:

> If the body image is intentional, the body schema, conversely:
>
> operate[s] below the level of self-referential intentionality. It involves a set of tacit performances – preconscious, subpersonal processes that play a dynamic role in governing posture and movement. In most cases, movement and maintenance of posture are accomplished by the *close to automatic* performances of a body schema …. The body-in-action tends to efface itself in most of its purposive activities.
>
> (Gallagher 2005: 26, emphasis in original)

The performances governed by the body schema are those acts of everyday physicality that we perform so frequently that it requires no conscious attention to enact them. There are whole-body movements or gestures performed with a limb that are so habitual and usual that we do not pay attention to the ways in which the physical motions are created and performed. Nevertheless, body-schematic processes can be 'precisely shaped by the intentional experience or goal-directed behavior of the subject. If I reach for a glass of water with the intention of drinking from it, my hand, completely outside my awareness, shapes itself in a precise way for picking up the glass' (Gallagher 2005: 26). Gallagher's definition of the body schema focuses on the disconnection of schematically controlled movements from intentionality (2005: 27). The intentional, conscious thought in the example above might be formulated as 'pick up the glass'. We do not think 'lift arm, extend arm, shape hand into x form'; under normal circumstances those instructions do not rise to the level of immediate awareness. In fact, 'such operations are always in excess of that of which I can be aware'.[1] Nevertheless, the body schema 'involves certain motor capacities, abilities and habits that both enable and constrain movement and the maintenance of posture' (Gallagher 2005: 24). What I wish to call attention to in the definition of the body schema is the automatic but nevertheless constrained quality of the movements governed by the schema. Although Gallagher focuses on the pre-consciousness of schematic movements themselves, I wish to examine the pre-conscious state of the form that those movements take. I agree with Gallagher that we do not consciously think to perform these acts, we simply do them; it is not only the movement that is automatic, though, it is the form that the movement takes. What I wish to examine moving forwards is how that form comes into being, how it is shaped and what pressures may be exerted on the form a performed movement takes as it evolves.

Body schemas include innate human physiological capacities for certain postures or movements (upright carriage, for instance) but are shaped from the earliest hours of life through imitation of others' bodily usages; Gallagher lays out the results of a series of experiments conducted on infant subjects (Meltzoff and Moore 1977 and 1983), which suggest that neonate imitation of facial gestures is possible from as early as one hour after birth (2005: 70). Significantly, experimental results suggest that infants are born with at least a minimal body schema, and that, while the 'knowledge of what I can do with my hands is *in my body*,' the performance of a specific physiological action improves with time and repetition to more closely match the observed performance (Gallagher 2005: 72, 74, emphasis in original). That is, our physicalization of habitual movement and gesture is moulded by our observation of others' performances of those same actions.

It is at this point in Gallagher's argument that I intervene with the formulation I define as the body map. I see the body map as the collection of social pressures that influence both the body image, which deals with portions of the body as they are understood by the conscious mind, and the body schema, which functions holistically and pre-cognitively. For Gallagher, it is *that* there is a cognitive schema directing our habitual movements; I ask the question of *how* that schema is pressured, moulded and shaped. Gallagher says, 'If throughout conscious experience there is a constant reference to one's own body, ... then that reference constitutes a structural feature of the phenomenal field of consciousness, part of a framework that is likely to determine or influence all other aspects of experience' (2005: 1–2). I ask how culture shapes lived, conscious experience in such a way as to landscape or develop the bodily terra. Schematic movements, per Gallagher, are movements we don't have to think about, but the features of that movement must be structured in some way and can be refined over time. I would like to move one level away from Gallagher's examinations: my question, then, is *why* those movements that we don't have to think about are accomplished as they are; why those specific, habitual forms and features of movement rather than all the other possible ways allowed by the purely physiological capacity of the human form for movement? If the body image and schema serve to construct embodied consciousness, what constructs the body image, and perhaps even more, how do those pressures soak into the body schema? That is, what structures the structure? My answer is the body map.

I view the body map as a structure that can be read from outside the body through reference to clothing and a broadly defined understanding of gesture, as well as a part of lived embodied experience. That is, while it may be possible to become or be made aware of features of the body map (as Gallagher asserts of the body image), it is not something that exists consciously in a holistic way. Rather, I understand the body map as a set of features to which the embodied subject unconsciously conforms. If body image is centrally connected to a personal attitude towards the body, I understand the historical body map as an attitudinally neutral construct, one that may be subject to politicized or critical readings but which is a reflection of cultural pressures exerted on the body image and body schema. The body map is thus focused, not on a subject's attitude towards her body, but on how that body's movement through the world is managed, pressured and structured in advance of conscious direction. While I borrow Gallagher's understanding of the body schema as a holistic entity created by the body's in-the-worldness, which is adopted by the body without conscious reflection, I also use the conceptual properties suggested by the word 'map' to localize specific body parts (topographical features, as it were) for analysis (Gallagher and Zahavi 1998: 145–6). The map, then, is both the large document and the individual features which make up the topography it graphs.

While I understand the cultural body map as being exhibited without the need for an individual's concentrated focus, I maintain that it includes behaviours knowingly adopted, if often unconsciously performed, including habits of social gesture which are purposeful acts, but are not gestures consciously guided through each repetition. In this category I place the physical habits of manners and decorum; for instance, the custom of gentlemen removing their hats indoors. That is to say, the body map includes features that have been added to the purely physical form; items, most especially clothing, that exist on or around the body and become part of its proprioceptive and kinaesthetic awareness.[2] According to Gallagher, 'The body schema functions in an integrated way with its environment, even to the extent that it frequently incorporates into itself certain objects – the hammer in the carpenter's hand, the feather in the woman's hat, and so forth – pieces of the environment that would not be considered part of one's body image' (2005: 37). I expand on this element of the body schema, graphing onto the features of the body map certain conditioned movements of social discourse; the learnt but unconsciously performed movements of bows or curtseys in any period would be one iteration of this element of the body map, but so, as I have said, would be the habit of removing a hat indoors. This feature of the body map includes not only the pattern of the performed gesture, but the impulse towards the gesture, the pre-conscious impulse of the body towards an action that has become reflexive under certain circumstances. The body map as I construct it includes features of habitual movement and gesture rooted by culture and the structure of clothing within the terra of the body schema.

How, then, do these features become embedded in the body? What are the mechanisms by which a body map permeates our embodied selves? Our bodies learn from the world they inhabit; a significant portion of our experience of our world comes via unconscious absorption of the way the world around us looks, so that we in our bodies are tacitly instructed by the bodies around us as an integral part of embodied existence, moulding our body maps. That is, by virtue of having bodies in the world, our bodies learn from the world they inhabit. Through observation, much of it entirely unconscious, we learn to stand, move, behave like the people around us, and they themselves draw upon culturally pervasive notions of the ideal in any given historical moment.

The significance of our unavoidable observation of the bodily practices of those around us is a central thesis of art historian Anne Hollander's *Seeing Through Clothes* (1993); she locates social habits of seeing as a major crux of embodiment and suggests ways in which a culturally pervasive body map may be created, acquired and transmitted. 'Bodily movement', Hollander argues, 'must always have tended, as it still does, to conform to mental self-images; and these must have been at least partly conceived with the help of external images' (1993: xi). This formulation has an obvious connection to the experiments in neonate imitation that Gallagher details, but Hollander's evaluation of Western art leads her to an analysis of how the structures of clothing may be imprinted on the body, on movement and gesture; this focus suggests ways in which the object being imitated may itself be created. She says that movements are made in accordance with a socially agreed-upon image of correct behaviour: 'This image cannot but include the look of clothes and an accepted sense of how one looks *in* them. ... [A man] will put his hands in his pockets, cross his legs, rest his arms on the back of a sofa – all "naturally," of course, but in imitation of an acceptable image of a man' (Hollander 1993: 315). My construction borrows from Gallagher's definition of the body schema the idea of pre-conscious movements, structured by imitation, and from Hollander's notion that clothing works to imprint certain habits of gesture on the body, both those bodies that are imitated and the subject body. I view the acquisition of a body map as a largely invisible process as strictures of polite behaviour, structures of clothing and saturation of visual imagery act upon the consciously adopted habits of dress and behaviour related to body image, and permeate the pre-conscious body schema. That is, while we may consciously adopt habits of dress and behaviour to support our attitudes and desires as they relate to body image, the pressures that shape those decisions are easily understood as training in manners or participation in fashion, rather than seen as value-laden social constructs. This understanding blends Gallagher's body image and body schema with the clothing-focused approach offered by Hollander to underpin my own understanding of the part of embodiment that I term the body map.

Because clothing exists in such a profoundly intimate relationship with the body, it is uniquely placed to enforce a historically specific body map by imprinting cultural expectations of appropriate bodily behaviour on the physical form. The fashions of the British middle and upper classes in the final twenty years of Victoria's reign and the reign of her son Edward seem to aim at remaking the living body in their own image, so that the shape of the clothing becomes the shape of the body.[3] The unyielding structures of clothing forced bodies into the moulded forms of garments; those garments and the body maps they enforced took part in the larger social understanding of gendered bodies and separate spheres. Men's bodies became defined by the three-piece suit, composed of trousers, vest and coat, worn over a shirt and necktie. This 'uniform' was a reflection of the shift from masculine prowess as demonstrated by strength of arms to masculinity and power defined by strength of mind, of industry, of pocketbook. It is the clothing of the businessman, not of the soldier; and the white-collared shirt[4] that was an integral part of the suit drew visual attention to the head and face, and helped to map the masculine form as one marked by a reliance on intellect and reason, a body that participated in the economic and intellectual commerce of the day.

Women's bodies, by contrast, were clothed and mapped in such a way as to emphasize their circumscribed sphere of activity. Gowns drew attention to a woman's hips, pelvic girdle and breasts rather than to her face. The constrictions of corsetry squeezed the waist while forcing bosom, abdomen and hips into fleshy billows above and below the narrowed middle. The corset itself was covered by a plethora of undergarments and garments, including camisoles, bodices and layers of petticoats and skirts. The hems of these skirts brushed the floor throughout the period, effectively pinning women to the ground. The Victorian idea of separate spheres permeated every facet of clothing. Women's clothes were composed of curves, conforming to the corset-imposed shapes of the body, while menswear created a male form defined by angularity and hardness; in women's dress, clothing created a body with rounded lines as its basis. In addition to moulding bodies to fit their structures, and creating a firm gender divide in the mapped body, garments also defined their wearers' social strata. The ability to change clothes multiple times in a day, especially for women, marked a body as both wealthy and socially engaged.

Through the consumption of goods and the production of a fashionable image, women were tacitly a means by which economic success was demonstrated to the world at large. The time it took to shop for fashionable clothing and to change ensembles multiple times a day, as well as the physical inactivity enforced by high-fashion gowns, demonstrated that a household was well-funded enough to allow the lady of the house leisure time devoted to the pursuit of fashion.[5] The trap of this monetary display was that it further constructed women as incapable of serious thought, interested only in material acquisition and the sensuous. A focus on woman as fashion plate called attention to a construction of gender that emphasized secondary visual or performed gender markers. A true woman must overtly appear to be a woman as late-Victorian society understood the gender; she must have long hair, wear skirts and evince feminine behavioural patterns, all, of course, culturally predetermined.

This brief examination of fashion in the late-Victorian period demonstrates how quickly an examination of dress expands to include examination and critique of the larger historical and social moment. It is in this dynamic that the examination of the body map becomes subject to specific critical lenses or to politicization. That is, to define the body map as both historically and culturally situated is to include much more than the conceptualization of anatomy and movement or the details of historical garment construction; it also encompasses beliefs about where bodies can be, what clothing they should wear or with what other bodies they may associate. To map a body, whether as an individual or a culture, is to construct it, to animate it and also to define its place within the social world of that culture. In the late nineteenth century, performers on the West End stage were particularly situated to communicate the pervasive body maps of the era's upper-class society.

The space of the realistic stage provided an ideal platform for the presentation and reiteration of the culturally normative and highly gendered body maps of the late-Victorian middle and upper classes. Leading West End actors and managers, often the same person with two roles, 'made themselves as middle class and utterly respectable as possible', Michael Booth notes, 'projecting an image that deliberately counteracted the (at morally best) raffish figure of the actor and manager in the popular consciousness' (1991: 23). The actor-manager also aimed to create a theatrical environment which made his audiences feel at home; in *The World Behind the Scenes*, Percy Fitzgerald describes Henry Irving's Lyceum Theatre: 'The whole has an air of drawing-room comfort. ... The spaces in front and behind the footlights seem to blend' (1881: 42). The late-Victorian period was the heyday of dramatic Realism, and the stage spaces in London's West End theatres were largely occupied by settings that mirrored the rooms of the upper-class homes to which patrons returned at the end of the evening. In addition, the dramaturgy of these plays was intimately concerned with the social interactions lived by the theatres' audience. Society comedies that made light of upper-crust foibles and Realist dramas engaged in critiquing the manners and mores that dominated the upper echelons of society required performance spaces as recognizable, minutely realized and authentic as true drawing rooms, dining rooms or conservatories. The same investment in realism pertained in costuming; actors and actresses were dressed for the stage by the same well-known and highly fashionable couturiers as their audiences. An Oscar Wilde might poke fun at Society's foibles, while an Ibsen anatomized their hypocrisies and failures, but in either case the world and the characters that appeared were closely observed and immediately recognizable facsimiles of real-life society. This investment in realistic costume and décor created a phenomenological space in which the bodies of actors onstage, dressed in contemporary clothing, could easily be read as synonymous with the bodies that inhabited the audience.[6] The play of the theatre space and the social theatre of the era were conflated in the construction of space and creation of the theatrical event (Booth 1991: 60). It was essential to this project that performers be able to present their audiences with bodies that were just as recognizable as the settings or characters.[7]

Actors exist at a social cell wall, able both to absorb the body maps visible around them, and to transmit them outwards, to model these maps for their audiences. They are uniquely positioned to mirror the 'natural' postures, gestures and movements of the culturally invested body map. The idea of a convincingly realistic performance is founded on the ability of a performer to adopt, not just the language and diction of any segment of society, but the body appropriate to the character played. It is my contention that in their performances, late-Victorian actresses in particular performed bodies structured by culturally pervasive maps that reflected gendered and classed expectations of physical comportment, carriage and behaviour. The modelling of the fashionable female body map on the West End stage was inescapably connected to the construction and constriction of high-end women's clothing. Actresses were literal fashion plates, modelling onstage garments designed and constructed by leading couturiers, but also modelling the movement and behaviour appropriate to bodies wearing those garments. Actresses who appeared in the fashionable forms of the day presented their audiences with templates for bodies. Significantly, these bodies were often defined as much by what they did *not* do as by what they did.

J. M. Barrie's *The Admirable Crichton* (1902) is a play deeply concerned with the effects of gender and class on bodies. The Ladies Mary, Catherine and Agatha Lazenby are daughters of privilege who deeply resent their father's insistence they serve tea to and make conversation with their servants once a month. They are even less enchanted when he announces that on the family's upcoming sea voyage, 'my daughters, instead of having one maid each, shall ... have but one maid between them' (Barrie 1918: 32–3). The ladies' immediate objections are firmly grounded in the ways their bodies are required to move and behave by the clothing of their class and sex:

Lady Mary	One maid among the three of us! *(Tragically.)* What's to be done!
Ernest	Pooh! You must do for yourselves – that's all.
Lady Mary	Do for ourselves – how can we know where our things are kept?
Lady Agatha *(rising)*	Are you aware that dresses button up the back? *(....)*
Lady Catherine	How are we to get into our boots and be prepared for the carriage? (Barrie 1918: 35)

It is not just that these women are unwilling to do for themselves, although Barrie makes it clear that that is so; they have been rendered incapable of doing so. An examination of a photograph from the original production, in the collection of the Victoria and Albert Museum, shows that all three of the actresses who played the Lazenby sisters were tightly corseted (Figure 4). If clothes made the man in the late-Victorian period, the corset was one of the primary ways of making women. Or, to be more specific, the corset was a way of making *ladies*. It was a major element of the period standard of decency for women, and it was the presence or absence of corsetry which constructed a woman as dressed or undressed, as presentable or *en deshabille*. As fashion historian Valerie Steele argues,

> The corset ... helped to create the appearance of a well-developed figure with a slender waist, which represented, for most people, the feminine physical ideal. By emphasizing the essentially female characteristics of the body, the corset functioned as a sexualizing device. Yet the corset was also widely perceived as *moral*; it was a necessity if a woman were to be decently dressed. To make a play on words, the straitlaced woman was not loose. (1985: 161)

The constriction of the corset as it influenced the body below it and the clothing worn over it shaped not just the torso, but also the manners of the body wearing it; in many ways, the corset imposed the body on the woman. The garment forced a woman to adopt a map of her body, her potential for movement, and her habitual interactions with the world in certain circumscribed, socially acceptable ways.

Lady Agatha's exasperated 'dresses button up the back' points clearly to the ways fashion imposes sets of physical behaviour on a body and forces the body into certain habitual ways of moving through the world by virtue of these imposed features of the body map. These women are actually incapable of dressing as their class requires, without help; it is impossible for their arms to reach the buttons at the centre of their backs. The impossibility of clothing themselves as their class and gender require carries forward into the sisters' general attitude towards any kind of physical exertion or work; when we first meet Lady Mary, she is complaining about how exhausting it has been to spend her afternoon shopping for engagement rings. The physical constrictions of corsetry and the back closures of their dresses, which make dressing themselves impossible, have translated into a broader mental construction for the Lazenby sisters. They have accepted a cultural body map that constructs a passive female form.

Lady Catherine's inquiry of how she and her sisters are supposed to get into their boots points to the same schema of restricted movement. High-fashion women's footwear of the period was no less complicated to don than couture-quality dresses. Securing women's boots required bending over (or propping a foot up) and either slipping each small button into its hole with a buttonhook or tugging the boot's laces tight around the leg. In any case, to

Figure 4 The original production of *The Admirable Crichton*, 1902. Lady Mary Lazenby (Irene Vanbrugh) is seated third from the viewer's right
© Victoria and Albert Museum, London

successfully fasten a late-nineteenth-century boot by herself, a woman would have needed to bend. Her hip joint would have needed sufficient mobility to allow her to bend forward or for her leg to lift, something the hip-length corsets of the period made both difficult and uncomfortable. Corsetry also made curving the spine in order to reach down along the body to a booted foot problematic. The steel or whalebone used to stiffen corsets resisted any attempt at a curvature of the spine; the job of the corset was instead to keep the female torso rigidly upright and unbending. As with dresses which buttoned up the back, in practical terms the constrictions imposed by corsetry meant that the fastenings of fashionable footwear were quite literally out of reach for the high-society women whose feet they graced.[8] The elaborate late-Victorian fashions worn by Society women were structured in a way that forced a construction of helplessness and decorative passivity onto their mapped bodies.

This effect, wherein the woman serves the same purpose as any of the picture frames or ceramics which can be seen in the photograph of the original production's Act One, is even more insistently underlined by the froth of skirt which billows at the hem of Irene Vanbrugh's gown. Vanbrugh, playing Lady Mary Lazenby, sits third from the right of the composition, looking out at the viewer. The lower skirt of her dress anchors the actress securely to the floor in a froth of dark ruffles; there is no hint of her feet, neither any indication below the knee that Lady Mary might have legs that allow her to move herself from place to place. This is one of the inescapable dynamics at work in the fashion for floor-length skirts that were a defining characteristic of upper-class women's dress throughout the late-Victorian period. These skirts visually anchored women to the floor and elided the presence of female legs; they also served as a practical encumbrance, especially as they were underpinned by multiple petticoats. The fragile layers of Mary's skirt act as swaddling, pressing against and tangling with her legs: they wrap and confine her lower limbs, preventing any stride larger than a restrained glide. Lady Mary seems to have been rendered effectively immobile by her skirts.

What I want to call attention to here is the way that women's long, full skirts obscured the structure and workings of their legs. In *Seeing Through Clothes*, Anne Hollander (1993: 339) observes, 'Exposing a woman's legs lays stress on her means of locomotion. ... Ever since antiquity [women's] clothes did a lot of their moving for them.' The garments worn by these women call attention to their immobility, rather than to their 'means of locomotion'; the yards of fabric and trim which the ladies wear imply a movement which their bodies are not permitted to substantiate. I argue that the constraints of clothing produced a cognitive paradigm in which a cinched-in, floor-bound body only understood itself as having the capacity to move within those confines; unconsciously, the body obeyed the instructions given by clothing. The visibility of the female body, and the way it can be remapped by a change in clothing and circumstance, becomes a part of the dramaturgical focus of Barrie's play; the plot of *The Admirable Crichton* provides the pretext for removing the women from the weight and drag of floor-length skirts with trailing flounces and ruffles.

The Admirable Crichton lays particular stress on the ways in which freeing the Lazenby sisters' legs from the weight and drag of skirts changes the way all three imagine the physical capabilities of their bodies. The ship upon which they sail is wrecked, and they find themselves stranded on a tropical island, following the lead of their one-time butler, the eponymous Crichton. The only one of the party with any practical skills, Crichton rises to the top of the island's aristocracy of talent. The froth and flutter of clothing which allow both the women themselves and the world around them to construct their bodies as decorative, assigned by class and gender to a passive role modelling the financial prowess of the men around them, disappears and is replaced for the duration of their island sojourn by physical exertion, relaxed clothing, and an accompanying empowerment. When the curtain rises on Act Three of the play, the castaways have been stranded for two years; the upending of social status has led to a social reorganization to what is 'natural' to the primitive island society (in marked contrast to the 'natural' order of things in overly civilized London). There is a parallel and radical shift from the bodies that were natural to the class and gender roles of London society to those constructed and remapped instead by the demands of island life.

At the end of the play's second act, Lady Mary is reduced to sitting on the beach weeping at the family's dependence on Crichton's practical skills. Her first entrance in Act Three presents an utterly changed young woman who uses her body very differently than she would ever have dreamt of doing in London: '*Lady Mary appears at the window. She is dressed in picturesque boy's garments of thin leather, feather leaves, etc., etc., and carries a bow and arrows, and has a slain buck and a couple of ducks. … [She] throws down the buck outside, and jumps in through the window*' (Barrie 1918: 84). Mary's new ensemble has freed her body for new kinds of movement. Without a corset immobilizing her torso or skirts controlling the movements of her legs, the woman who was once incapable of lifting a hand to display her engagement ring without the greatest *ennui* now thinks nothing of hauling game about and leaping freely through a window.

Her new clothing has made Mary's wider range of motion possible, but the clothes in and of themselves do not guarantee a change in movement. What they do is provide a new set of information to Mary about what her body is and how it can work, allowing her to conceptualize its structure and potential for physical action in broader terms than the clinging and trailing gowns of London did. In *The Admirable Crichton*, Barrie's themes are writ large on the bodies of the characters he presents. We are shown the embodiment of feminine freedom from constraining corsets and skirts, and left in no doubt that the released woman is happier for it. Soon after her entrance, the family's tweeny maid asks Mary, 'Ain't you tired?' 'Tired', Mary replies, 'It was gorgeous.' (Barrie 1918: 85). The laziest of the girls has shed the map which constructed her body as immobilized and waited-upon in favour of a mapping which embraces physical uses which would have shocked her Act One self.

The castaways are rescued from their island, though, and in Act Four, we find the ladies returned to London, and to their corsets, skirts and aristocratic lethargy. With the shift back to their previous modes of clothing and behaviour comes a return to the original bodies we saw the sisters inhabit. This second remapping causes Mary, in particular, more difficulty than the first, although she is asking herself to return to the long-held physical patterns which governed the majority of her life. She can no longer imagine the body she is socially expected to have; the map that culture requires cannot be the embodiment of the woman Mary has become over the course of the play. In the aristocratic London milieu to which Mary, Catherine and Agatha are accustomed, they are being prepared for marriage to another member of their class and society so that they can become the mothers of the next generation of that very class. Beyond this, they are expected to be passive reflections of the masculine labour and wealth that underpin the society of which they are members. Their body maps, therefore, are structured by this sociocultural expectation of decorative femininity. Barrie hints, however, that perhaps the body schema has been remapped in more permanent ways than her newly readopted body image might suggest: Mary has trouble remembering not to cross her legs at the knee. Is it possible that her time on the island has allowed a less-restrained body map to shape the body schema to the extent that what was once pre-conscious, 'natural' femininity must now be a performed body image that no longer conforms to Mary's most deeply embedded schematic instincts?

With Mary's ultimate reassimilation into London society, Barrie disavows the emancipated bodies he briefly granted the women of *The Admirable Crichton*. Mary accepts that, in London, she cannot attach herself to Crichton but must resume her engagement to the priggish Lord Brocklehurst; she must therefore make conscious the performance of the socially requisite female body which she once used without thought. That is, Mary is forced to jettison her new, freer body map and remember the features her body is required to adopt in 'civilized' London. Although the play flirts with the notion of body maps which liberate the female form, in the final instance, it is the standard Victorian notion of the female body which is affirmed by this presentation of women's imaginative mappings.

If, as Gallagher says, 'our human capacities for perception and behavior have already been shaped by our movement' (2005: 1, 3), it becomes a matter of singular importance to parse the forces that themselves shape that movement. Movement structures and defines not just our embodied experience of the world, but 'the human experience of self', the understood *potentials* for engagement with the world, where movement is understood in both its literal, physical sense and in its metaphorical sense of flow within and between genders, classes and social groupings.

When they appeared in high-fashion clothing as characters drawn from the very recognizable contemporary world, actors and actresses embodied society's expectations of the body. These expectations were directly connected to the perpetuation of a social order designed to underpin the British imperial project; distinctions between gendered bodies in the late-Victorian period aided in a colonization of the body akin to the contemporaneous colonization of geographical territory. This project depended upon a male body mapped as intellectually vigorous and physically stalwart and resolute. The *de facto* uniform of suit and high-collared white shirt created the male body as a column of strength. This body was situated at the top of a Darwinist social structure; the mental prowess of the British businessman and politician led the nation. These same bodies were positioned at the head of the domestic social structure as well. The strength of the male body, the paterfamilias, was opposed to the construction of a female body that was mapped by the confines of clothing. Women's garments, especially the corset and floor-length skirts, created a female body map that equated the feminine with mental weakness and childishness, while at the same time showcasing the sexual characteristics that made possible the continuation of imperial society.

My contention is that body maps are the pattern by which society imposes its values on our bodies; maps contain strictures of movement, carriage and gesture which are imprinted upon both body image and body schema, shaping our habitual movements. Body maps are the culturally imposed and heavily value-laden parameters for conscious behaviours that relate to constructing and maintaining a desired body image; these behaviours then impact schematic processes related to aspects of kinaesthesia, proprioception, aspects of movement and autonomic movements. Embodied in the mapped form, then, is the iteration of the culture which produced the map. It is thus possible for cultural assumptions to inscribe themselves on the body map in such a way as to become larger structural understandings of how the world works. So, for example, a floor-length woman's skirt connects her always to the ground and helps to map a physically restrained body; if this mapping transmutes into a larger schematic structure, it easily becomes an embodied understanding of women as homebound and domestic. The further danger of transitively understanding 'this body in this moment' as standing for 'all bodies like this always' is the danger of universalizing specific mapped bodies into a foundation of metaphorical schemata; the tendency to understand 'abstract' as 'eternal'. The quotidian world can be seen to change regularly, but the abstract, conceptual world has historically been understood as a realm of first principles where Truth resides. 'Structures of our bodily experience work their way up into abstract meanings and patterns of inference,' says Mark Johnson. His work points to the dangers (recognized by post-modernity) posed by the abstraction of the acculturated body map, which contains the embodied expression of heavily weighted cultural values, assumptions and mores (1987: xix).

The liminal status of theatrical performers in the Realistic productions of the late-Victorian period creates an observational and experiential space in which the normative body maps of the period were absorbed by Society. The body map that I have plotted here is a landscape produced by the social dictates of manners, decorum and gesture within polite society, and further imprinted by the structures of clothing. This accumulation of influences, I suggest, trains the body to unconsciously move and behave in certain habitual ways, which themselves exert pressures on the composition of the body schema, as defined by Gallagher. That is, the pressures reflected in the body map train the body to behave *in this way*. The constant and consistent performances of these trained bodies reinforce (or potentially subvert) the ways in which historically specific power structures were reinforced as concrete physical pressures are transmuted into overarching structures of understanding. To chart the features of a body map, then, is to understand the cultural landscape writ on the body, to define the means whereby cultural power structures are embedded in embodied consciousness.

Beyond what I've discussed here as a reading of one particular body map, I'd like to suggest that there is a way of understanding elements of the theatrical event that is implicit in the case study I've presented. A central part of theatre is the seeing of it, the visual absorption of the staged event; in a very intentional and focused way, as audience members we are engaged in the visual consumption of bodies. Clothed bodies. Performing clothed bodies. Theatre brings together the clothed body with visual attention in a very different way than the observation that is a part of daily life. On a stage, these bodies become a mechanism of social education that is hidden in plain sight. What the theatrical event transmits to an audience is more than the message of the text (or non-text) being performed, more than the themes or morals of the dialogue. The entire structure of the event carries weighted meaning. This dynamic has been acknowledged as it regards issues of diversity; that is, many theatres make a concerted effort to present diverse casts with the understanding that it is beneficial to reflect a multicoloured, multi-gendered, multi-abled society. Presenting these diverse ensembles from the stage does more than help to normalize an understanding of society that includes a multifaceted humanity, it presents that diversity as *already* normative. Phenomenology has done much the same thing for the understanding of the weighted construction inherent in the *mise-en-scene* presented onstage. I would like to offer the body map as one way of understanding the value-laden tacit communication in which clothed, performing, onstage/staged bodies engage. While the analysis presented here is not and cannot be multicultural or apply directly to multiple time periods, I think it is possible to transfer the idea of the body map.

What might be revealed by extending our exploration of the transmission of information from clothing to the surrounded and surrounding bodies via the theatrical event? What if we examined, for example, the court dress handed down by patrons and worn by Elizabethan actors, just as a brief thought exercise? We might note its stiffness, the padding or corsetry, the ruffs, the tightly fit garments that constrain the body to the forms of social gesture required by the highly structured and stylized life of the court. These are bodies that do not move without reference to the monarch, as is clear in many of Shakespeare's court scenes. This formulaic stiffness imposed by clothing on the body is one element of the systemic control exercised over courtiers by monarchs in Shakespeare's plays, and more broadly, by Elizabeth I. The clothing becomes one method by which the message 'I control your body, in all its motions' was transmitted. This control extended beyond physical gesture to presence or absence at court, to permission to marry, and finally to life or death.[9] To analyse the theatrical event through the prism of the body map is then to examine the event as a constituent element in the creation of cultural embodiment.

In this chapter, I've looked at a particular body map at a particular moment performed on a particularly located stage. What may remain unsaid in that examination is the idea that we perform our bodies differently after seeing bodies performed. Bodies, precisely because they are so constantly and intimately ours, our prism of experience, are always familiar.[10] By presenting bodies to us, the theatrical experience to a certain extent makes the familiar strange by placing it as an object for consumption. The theatrical event calls attention to the body clothed and moving through space, thereby presenting it to the audience to be read. I do not suggest that this reading happens consciously, but that the privileged space of the stage has functioned as a place for institutionalized, sanctioned watching. We watch these onstage bodies, read them and are impacted by them without our conscious notice. And in this silent process whereby the social constructions, the body maps, communicated by these onstage bodies as they wear clothes and move through the world have been read and understood by our own bodies, the fact of the communication becomes tacit. The cans and can'ts and wills and won'ts of our bodies, which are part of systemic culture, become invisible. What I hope the body map does, finally, is to provide an analytic space in which these pressures on embodiment can be made manifest.

Another Way of Looking:
Reflexive Technologies and How They Change the World

Matt Hayler

I want to try something. I want you to sit with this book and a computer, or a smartphone or tablet, any screen where you have access to the internet and can watch a video and search for some pictures. I'm going to ask you to look at some things, to mix media with me, like we might break bread and talk, or gather around a laptop at a friend's house. Part of what I want to discuss here relates to the effect of my being able to ask you to do this, to send you somewhere else and suddenly, instantly, to have a shared knowledge and a context for discussion, for us both to have something in mind. I want to talk about a variety of performances after digital technology, most broadly about how we perform our daily tasks differently after looking at things differently, and the ways in which our devices enable this to occur, more than ever. I want to look at a series of contemporary examples of movement and action and representation which, while inconclusive in themselves, may function as early symptoms of a shift in a distinctive technological effect. While this effect isn't new, it is certainly heightened in the age of digital distribution, and, as I'll go on to discuss at the end of this chapter, it may underpin a New Aesthetic in our cybercultural landscape.

I want to build here on Don Ihde's post-phenomenological distinctions between types of technological interaction and to offer a new category which furthers the post-phenomenological aim of exploring how humans encounter a world that is mediated and co-constituted by our engagement with material artefacts. Technologies, in short, don't just allow us to do new things, they enable us to see new things, and to become new things – we are not human without our tools, and our tools seem to always be changing, as do we. What does the digital bring to our visual grammar, and what does a technology have to do to impact upon the rules of our comprehension?

Post-phenomenology and technological deployments

Don Ihde's American post-phenomenology is an updating and nuancing of phenomenological concerns, particularly with regard to the role that technology plays in human action and perception. The field tends to be case-study and praxis focused, intently interested in how humans actually deploy artefacts in daily life and in special cases, and combines elements of pragmatism and classical phenomenology to form an approach sensitive to embodiment, social forces and the interactions between humans and their tools. As Ihde described his approach, 'By focussing upon embodiment and inter-relationality, [postphenomenology moves] away from the separation of the machinic and the human to the interaction of the machinic and the human' (2010: 46). It is this sense of interrelation that led Ihde to document the ways in which humans are influenced by their technologies, and to develop a taxonomy of these deep relations.

In *Technology and the Lifeworld*, Ihde outlines four types of encounter with artefacts.[1] The first, 'embodiment relationships', describes the ways in which users bring technologies 'onboard' and treat them as a part of themselves, a process that I've described elsewhere as 'incorporation'.[2] Examples of such might include a tennis player's use of a racquet, a guitarist's guitar, a driver's car. In each of these instances, the technology 'melts away', becomes subordinate to the work to be done[3] – the user treats the tool as an extension of her body.[4] Embodiment relationships shift our conception of what our bodies can achieve and make our boundaries seem more porous.

The second form of interaction that Ihde describes is a 'hermeneutic relationship'. These encounters see users looking 'through' artefacts in order to interpret the world. A readout from a machine, for instance, marks the hermeneutic observation of a phenomenon in this sense, bringing the universe, or bacteria or the quality of steel 'closer' (or available at all) through a telescope, microscope or spectrometer. Again, our sense of the achievable, the perceptible, is altered and we tend to think of our direct access to the world as being expanded even as such hermeneutic technologies must *condition* our access even as they enable it at all.

The third category comprises 'alterity relationships', where users consciously focus on an artefact as an embodied 'other'. Drawing money out of an ATM, for instance, requires that we focus on the artefact itself in order to act – we do not look through the device, nor treat it as a part of ourselves; its use is in its otherness. Microwaves, kettles and automated checkouts very clearly provide new potentials outside of ourselves.

Finally Ihde describes 'background relationships', the deployment of artefacts wholly outside of our attention even while they retain their ability to shape or flavour our experience. Such background technologies tend to impact upon our lived environments such as the presence of thermostatically controlled heating.

In establishing this language, Ihde investigates the ways in which the use of artefacts affects (and effects) perception and action. Embodiment relationships, for instance, change our sense of the limits of our bodies and our capabilities. Hermeneutic relationships tend not to interfere with our boundaries, but certainly change our conception of the possible, the visible, the interpretable. Alterity relationships are less entwined with our being, but they can still be profound, co-shaping a world of money and food and entertainment made available with a few small gestures. Background relationships are multiplying and becoming increasingly subtle as people increasingly monitor their surroundings and cede control to machines, moving beyond ambient temperature control to stock market manipulation, navigation, health checking, information searching, recording and documenting, and more. The essential point from each of Ihde's technological potentials, then, is that our artefacts can grant us a new outlook on the world even as they change how we conceive of our place within it.

Ihde offers us an astronomical example: 'A hand-held telescope magnifies both the Moon and our bodily motion and thus makes it hard to maintain a fixed focus upon our observed heavenly body. Here is yet another clue to the complexity of embodiment: every change in our newly magnified world is also a change in our embodied experience' (2010: 58). This combination of an embodiment and a hermeneutic relationship with the telescope, an extension of the body and a looking through the technology, results in our feeling the slightest movement as magnified a thousand- or ten-thousand-fold; to make a minute gesture vaster than our natural satellite is to feel the distance between us profoundly. In this way, post-phenomenology calls for us to think of technologies as far more than implements which exist outside of ourselves and to which we must turn in order to perform a task. Instead our artefacts alter how we understand our physicality and the set of default practices that we consider ourselves able to achieve, even in their current absence. We could not travel at speed without the car in the garage or the plane in the hangar; we could not type and print without the computer in the next room; we could not hunt effectively without the spears in the rack nor butcher the catch without the knives near the hearth, and we cannot even *intend* to do so without these items. As the archaeologist Timothy Taylor describes this entailment, 'It is not too much philosophy to say that the emergence of technology was and is intimately connected with the extension of the range of human intentionality. ... The existence of [artifacts] not just allows actions but suggests them.' The central post-phenomenological insight is that technologies both prompt and structure our engagements with the world.

Grammar

I want to offer a new technological relationship that fits into Ihde's post-phenomenological continuum and marks another distinctive way in which technologies can be deployed and impact upon their users' experience. I'm interested, here, in those interactions with artefacts that alter our abilities by manipulating the *grammars* through which we meaningfully conceive of our potential for action. Such interactions focus on the visual replay of the body in motion, controlled by the user or an assistant, or the viewing of some other body in a medium, in photography, cinema or videogames, for example. The mirror and the manipulated camera will be the totemic technologies for what I want to call, after Ihde, 'reflexive relationships'.

I've chosen the term 'reflexive' for its various definitions, the relevance of which will, I hope, become increasingly clear. Reflexivity describes circles of cause and effect, feedback loops, the self's relation to the self, individuals to individuals, both a turning inwards and an iterating outwards. There's also the link to doing something by reflex, and this training into the user's being is also important. So what is a reflexive technological relationship? And what does it do? We need to start with grammar.

Our experiences don't exist in a vacuum – experience never goes unstructured. That a communal language conditions our encounters with the world, for instance, is at the heart of the critical theory of the mid-late twentieth century, from the influence of Ferdinand de Saussure's radical revisions of semiotics to Michel Foucault's discussions of linguistic power and ideology; from Jacques Derrida's critique of logocentrism to Judith Butler's description of the role of language in performative gender roles. Each of these thinkers, and the reams of theory and philosophy that they have influenced, are simply recent examples of a long history of thought about the power of language to condition our reception of the world. Ihde draws on this history in his phenomenology of multistable visual effects,[5] alongside the 'hermeneutic phenomenology' of Heidegger and Paul Ricouer,[6] in order to describe the ways in which raw sense perception is primed by the stories drilled into a prehending subject by her prior experience:

> In a hermeneutic strategy, stories ... are used to create an immediate
> ... context; they derive their power of suggestion from familiarity or
> from elements of ordinary experience. The story creates a condition

that immediately sediments the perceptual possibility. In untheoretical contexts, this has long been used to let someone see something. Storytellers, mythmakers, novelists, artists, and poets have all used similar means to let something be seen. Plato, at the rise of classical philosophy, often paired a myth or fable with argument or dialectic. Within the context of the story, experience takes shape. (2010: 61)

Elsewhere, Ihde states the case more plainly: 'Perception takes shape within and from the power of suggestion of a language-game. It sees according to language. This strategy is the basis of what has become known as hermeneutic phenomenology' (2010: 61). Now, I don't want to make this point too strongly, to suggest that language plays the dominant or even one of the most significant roles in how we encounter familiar and novel objects, to over-dramatize the linguistic turn. But language and storytelling does represent a clear way in which how we perceive and conceive of the world can be affected by the sediment of previous experience. In this regard, language can act as a repository for our knowledge and a primer and shaper of future action in both explicit and subtle ways.[7]

Several recent experimental programmes have rekindled interest in the explicit role that language can play in perception, an area of research critiqued to the point of near extinction in the last years of the twentieth century. Landau et al.'s 'The Influence of Language on Perception: Listening to Sentences about Faces Affects the Perception of Faces' (2010), for instance, demonstrates what its title suggests, as does Gary Lupyan and Emily Ward's 'Language Can Boost Otherwise Unseen Objects into Visual Awareness' (2013).[8] But my acceptance that language, language-games and storytelling can structure perception shouldn't be mistaken for a full commitment to (what has become known as) the Sapir-Whorf hypothesis, strong linguistic relativism or wholly social constructivist or constructionist models. Rather, I intend a weaker notion that goes beyond linguistic determinism to suggest that we take on a wide variety of strategies, often metaphorical, never simply or solely linguistic, that structure the ways in which we respond to the world; language and stories play an important part, but no one strategy consistently dominates across encounters. That the stories culture tells us have an effect is an intuitive claim, however: I'm primed to find dogs companionable, words in books important and to consider enduring monogamous love achievable and desirable, at least in part, because of my cultural background. All of these things could no doubt manifest independently of my cultural priming, but equally undoubtedly my beliefs, perceptions and investments of effort are also deeply affected by my linguistic and cultural conditioning. Such concerns with language, media and storytelling are familiar to the humanities, if not always in quite these forms.

Language and stories matter both for our imagination and for our perception of things, but the repetition of experiences more broadly builds up a 'grammar', or sets of grammars, that we can deploy in order to structure future situations. And grammar seems the right word: the fairly intuitive case of language detailed above gives us a model for the broader understanding of a grammar as a set of rules that we tend not to intentionally learn for our mother tongue, the rules that often baffle us when we either pursue a new language or try and explain our own to others. A linguistic grammar, and all of its effects, is typically deployed without thinking in both expression and comprehension, shaping our thoughts and whittling possibilities down to something more manageable – the meaning of a word, as we know, is to a large extent determined by the rules of the game which surrounds it.

I'm interested here, however, in the kinds of grammars which might emerge out of our repeated encounters with the visual arts and other technologies of representation. An effect of visual grammar can be seen, for instance, in montage. Our training by film, photography, painting and particularly comics and other sequential art has developed our sensitivity to the drawing of connections between consecutive images, to tell a story or to produce a sensation by affinity. Imagine three detailed paintings; take a couple of seconds to visualize each one in turn. First: a baby's head. Second: a chicken's egg. Third: a hammer. Picture this triptych and you should start to feel uncomfortable, because you start to turn the images into a narrative. There's no inherent reason to do that, but we have been trained to make metaphorical and temporal links between images. There is no story, and yet we somehow fear for this Platonic child out there somewhere due to our media training colonizing our thought and affecting our perception. The grammar affects our reception of each unit and the whole without ever changing their appearance – perception, then, is more that it looks; it is always conditioned by more than we can consciously consider, always partly a product of more or less apparent choices and influences.

Similarly, if I say, 'imagine Antarctica', then most readers will now have a very clear image in their heads: snow and ice, a white desert, maybe some penguins or hardy seals, possibly a shape on a map or globe. And I'm guessing that you probably haven't been there, and yet that image, for all that it's just a strange composite of the hundreds and thousands of images that you've seen, has a real effect on your understanding of the earth; visual media and the stories that we are encouraged to construct, invade our sense of our lived environment. Susan Sontag, in *On Photography*, noted that we can no longer look at a sunset without what's really in front of us being judged in relation to the sunsets of the films, photographs and paintings that occupy our minds as much or more than our real prior experiences of the phenomenon – 'Certain glories of nature … have been all but abandoned to the indefatigable attentions of amateur camera buffs. The image-surfeited are likely to find sunsets corny; they now look, alas, too much like photographs' (1977: 85). Sontag was concerned with how reality becomes distorted by the ways in which we saturate ourselves with images and their particular cultural valence, but that's also a significant feature of how perception functions: the marshalling of past experience always enables the production of a useful distortion and reduction of the immense totality of sense impressions that we access every instant of our lives. Visual grammars, too, provide us with vital and flexible contexts for navigating encounters that are always at once both too rich and isolated – grammars simplify and they draw bonds between instances.

My last brief example requires you to head to the internet: an image search for 'tilt shift photography' should bring up plenty of sample pictures of the photographic effect that I want to discuss. If you haven't encountered the phenomenon before, try and work out what's strange about these images.

Some readers will think straight away 'oh, they're models', and on reading that I suspect that more readers might see that this is the case. The distinctive blurring, the shallow depth of field, is familiar from photographs of miniatures. Another image search (ideally in a new tab for comparison) for 'slinkachu' will provide plenty of examples of these features demonstrated with the photographer's signature images of tiny plastic figures living and working in a macro world – again, note the blurring that tends to occur at the top and bottom of the image, an artefact of the camera's apparatus as it focuses on the fine details of small subjects. If you return from Slinkachu's work[9] to the search results of 'tilt shift photography', however, you may start to notice that, despite the similarities between the image results, you're not, in fact, looking at the same kinds of subject. What has become known as the tilt-shift effect[10] is in fact an *emulation* of the visual artefacts that Slinkachu's work includes because of the nature of its subjects – the vast majority of the 'tilt shift photography' results are of macro objects made to *appear* as miniatures, not miniatures that mimic life-size subjects. Again, we have been trained (or, if the effect is new to you, then you are being trained) to see selective focus and fore- and background blurring as indicating, as being the *unavoidable* requirement of capturing images of small and finely detailed subjects – a person alive before this kind of cultural conditioning, or outside of its exposure, simply cannot conceive of these things in this way. The images and the subjects depicted remain the same, but the learning of a new grammar, or at least of some grammatical rules, changes our reception, conception and potential for intention; the grammar makes the world appear as differently as the apparatus of the camera or the post-production effect.

Reflexivity

So these are some examples of how typically subconscious grammatical rules might affect our perception, changing the objects that we encounter and our sense of what might be achievable with them. They are also subtle forms of the reflexive relationships that I want to describe. A reflexive technological interaction is one where an artefact causes our conception of our potential for action to change by altering the rules of our visual grammars. As Ihde's four relationships shape the user's world, so can reflexive relationships be seen as similarly co-constituting.

The brief examples above had clear grammatical effects, but subtle reflexive impact. I now want to spend a little more time with three dance performances that have, I want to argue, been produced by users whose embodied grammars (i.e. their contextual rules for conceiving of movement and performance) must have been conditioned by a culture of reflexive relationships with the technologies of (at least) mirrors, cinema, video cameras and photography. In these examples the reflexive effects are clearer, but the grammatical shifts, the rules, are more nebulous than in the examples given above.

So, another fresh tab please, and a search for 'amazing tron dance by wrecking crew orchestra'. This video (Rushgaroth 2012) is a stage performance by the Japanese dance group Wrecking Crew Orchestra. In the clip, each member of the group has strip lighting attached to their costumes that can be triggered by remote or by timing devices so that on a pitch-black stage we seem to be watching wireframe figures, 2D representations of dancers that appear and disappear as required, flitting in and out of being. This is a performance that is indebted to digitization in a number of ways. Most obviously, the programming of the lights, and the wireless connection that is required to trigger them, is a particular product of this moment. This performance simply couldn't have been achieved, at least not with this freedom of movement, even ten years ago. Second, there's a clear influence in the costuming from neon signs, wireframe computer graphics and, more specifically, the 1982-film *TRON*,[11] a cult classic with a recently made sequel that re-popularized its aesthetic. The original movie was predominantly set inside a computer simulation, and was one of the first films (and certainly the most popular) to extensively use computer-generated sets; it is certainly a landmark change in both film aesthetic and the potential for the representation of both videogames and virtual environments on screen.

But besides these requirements and influences, technology and digital culture are felt in the *choreography* of the Wrecking Crew Orchestra performance, in the movements undertaken by the dancers and the logic of the triggered lighting that brings them into view. This is a dance sequence that couldn't only not have been produced before, but couldn't even have been *conceived of* in terms of the effects that it deploys because the motivation for those effects emerge from the cultural baggage of over forty years of digitally tampered-with visual imagery and 200 years of photographic experiments.

Most subtle, perhaps, is the way that the dancers tend to appear on the stage as their costumes are triggered, either simply bursting into existence or, as seen clearly at 1:35–1:37 minutes, flickering into being. Both of these feel like the 'right' way for digitized characters to occur; not as ghostly presences, simultaneously here and not, but binary, off then on, with any glitchy coming 'online' being an oscillation between, not an overlap of the two states. Digital technologies have suggested ways of moving to the dancers, and ways of choreographing the lighting; they suggest ways of acting to the actors, but also ways of receiving to the audience – the sharing of a common grammar of digital imagery conditions the way that the performance plays out and is contextualized, and in this sense it co-determines its existence.

Peter-Paul Verbeek's take on Ihde's post-phenomenology, in *What Things Do*, is useful to my understanding of 'co-determines' here. Verbeek focuses on the ways in which humans and their artefacts construct the encountered world together through technological mediation:

> Formulations in terms of the 'access to reality' offered by an artifact should be read as relating to the way in which an artifact makes possible the constitution of the word in the very process of perception. Humans and the world they experience are the *products* of technological mediation, and not just the poles between which the mediation itself plays out. (2005: 130)

We can think of Ihde's four relationships as a taxonomy of distinctive types of mediation in this sense – artefacts don't just get between humans and the world, they structure the nature of the world's appearance. My suggestion of the category of reflexive technologies fits as another kind of post-phenomenological mediation, a co-shaping of the experience of aspects of the world as lived, as received through trained grammars. Grammars, in this understanding, function like mediating technologies (and the reflexive relationship marks their connection), and Verbeek's description of mediation remains applicable: 'The transformation of perception has, according to Ihde, a definite structure involving amplification and reduction. Mediation always strengthens specific aspects of the reality perceived and weakens others' (2005: 131). As we've seen, so do grammars.[12]

Another example, more closely allied to the impact of the tilt shift grammatical rules that I described above, can be seen in the Wrecking Crew Orchestra video at 1:00–1:14 minutes, 1:45–1:51 minutes and, most clearly, at 1:57–2:02 minutes. At 1:00 minute, the dancers arrange themselves so that an action seems to carry across several bodies; at 1:45 minutes, the flickering effect is combined with this new effect; and at 1:57 minutes, two dancers fight like videogame characters until a landed punch sends one combatant reeling backwards, with his fall transferred across the frozen poses of multiple other dancers via the same rapid switch between lighting rigs. And we've seen this kind of thing before, even if we're not sure what *one* thing is being referenced; another photographic effect is being quoted here.

The overlaying of images (an image search for 'sequence photography' will bring up some great examples of the effect) has become a staple of sports photography and some particularly kinetic advertising campaigns. It's been around for a long time in various forms (we'll return to a pre-digital example shortly), but its frequent use, and particularly its use in moving images, is much more recent, again largely dependent on the new digital camera technology. Who knows how consciously the choreography of the Wrecking Crew Video drew on the sequence photography effect, but there seems, undoubtedly, to be a close relationship here. Sequence photography has become another rule in our visual grammar: a wholly unnatural way of looking at the world that's been made natural, an available and intelligible way of capturing and depicting chronology that builds on the excitement of still photography's isolation of the fleeting and cinema's revelling in kinetic motion. I'll return to the implications of this after setting up another kind of dance performance as further evidence.

The next videos that I'd like you to consider can be found by searching 'pumped up kicks dubstep' (WHZGUD 2012) and 'take me away dubstep' (WHZGUD2 2011).[13] Again these are digital cultural products in several ways. Most obviously they're meant to be YouTube clips and they look pretty good despite being filmed on an affordable home camera. The 'PUMPED UP KICKS|DUBSTEP' clip has also been watched over 100 million times, making an essentially free-to-produce-and-distribute performance hugely impactful in terms of its reach and potential to inspire others. But I'd like to focus on how, again, Nonstop exploits particular aspects of contemporary visual grammar as both inspiration and interpretative framework for his performance. First, there are the 'popping', 'locking', 'waving', 'botting' and 'tutting' movements that are present throughout both of these videos,[14] and although each of these styles are constantly updated as they are passed down by generations of dancers, VHS, DVD and now streaming-service performances have been vital in maintaining what is conceivable. These are not natural movements; it would be quite possible for a particular region or neighbourhood to not have a member of their generation of dancers able to accomplish a particular movement to a high standard due to the requirements of both the intensity of practice and pre-existing skeleto-musculature. The video archive, therefore, is essential to both growing and maintaining these kinds of street dance techniques, particularly in small communities over the ebb and flow of dance styles' popularity.

What is also ever-present in these performances from Nonstop is the clear influence of slow motion and the manipulation of the unfolding of captured motion through the apparatus of first film and now digital editing (this is perhaps most clearly expressed in the 'PUMPED UP KICKS|DUBSTEP' clip at 1:03–1:20 minutes where Nonstop's slow-motion progress forward then appears to be kicked into reverse). The effect of seeing human movement slowed down, repeated and 'stuttered' without technological intervention, and the rapid transition between various perceived speeds, is what gives these performances some of their uncanny edge. As we might initially wonder whether they have been doctored (neither have been), and then marvel at their authenticity, we can see another change in the conception of what the body in motion *can* look like. Nonstop makes what has previously been (at least for most of us) the province of the editing suite available, at least in theory, to every owner of a body – we can look, act and be differently.

These dance videos are haunted by a history of images being slowed down in order to amaze and to teach. And pedagogy in dance has always deployed reflexive technological relationships in order to produce its effects: the first must surely have been observation and imitation, watching the pleasant shapes of others' actions and trying to replicate them in our own bodies, listening to the corrections of impartial observers, always trying to judge from a viewpoint outside of our ancient selves. The mirror, if the term 'technology' must be tied to an artefact, is the other candidate for the first reflexive technology for dance, allowing the dancer to see herself, for the first time, from the perspective of an observer. The second great leap would have been multiple reflective surfaces allowing views of the rear and side aspects of the body in motion – such reflexive technologies would have changed performers' abilities to conceive of themselves in motion (as they must have changed the human relationship with the self more broadly); the understanding that a particular proprioceptive or kinaesthetic *feel* looks like *this* from within, but like *this* from the front and *this* from the side and back is again to alter the structuring context of future experiences. Similarly, video cameras now allow for both the capturing of the self for later review, but also the capturing of others on the other side of the world for contemplation, dramatically expanding the palette of embodied expressions available to every beginner dancer or expert choreographer. Digital editing techniques, of which sequence photography is a minor example, only take this further, provoking the potential for the imitation or, at least, the influence of a new kind of representational media.

Perhaps the most significant recent touchstone in this regard, something that re-energized what could be shown on screen by taking and exploring the time to observe, was the original *The Matrix* movie's use of 'bullet-time' (1999). A last video to remind readers of the effect can be found by searching 'the matrix bullet time helipad fight scene' (actionjawa 2010). First we see Neo, the protagonist, shoot into the camera, directly at us, in what is very clearly real time; it's an intense perspective to have on a pair of fired pistols and this heightens our attention on the motions, rendering them simultaneously alien, but also incredibly familiar from action cinema. The reverse shot sees the agent, a nameless antagonist, dodging Neo's bullets; he disrupts our expectations of what an enemy target 'should' do as he moves, to us the audience and to Neo, whose viewpoint we now occupy, impossibly quickly. It looks ugly. The animation is a strange combination of organic and digital noise, and I was never really persuaded by the effect as a depiction of speed. But then we see the reverse shots: Neo runs out of bullets and calls for help, but the agent steps up, places the barrel of the gun into the camera, and fires – the intensity is raised again and from the image and the soundtrack we know that we're still in real time, re-emphasizing the speed of the agent's evasive movements. We expect the next shot to be the mirror of the second shot, where we would now take the agent's-eye view of events and see the path, and the target of his bullets. But instead we stand behind Neo, hear the soundtrack slow and enter into Neo's experience of time. And it's beautiful. We're granted a godlike control, able to move around him, not stopping time, but bending it to allow us to intently observe, a whole new way of looking at the world, like turning something around in our hands in order to look at how its facets glint. We take Neo's newly discovered powers and use them to our own ends to observe him from a vantage that's as denied to him as that of any other performer reliant on imitation, mirrors and cameras after the fact – instead of his perspective, as viewers we adopt his perception and put it to work.

This way of looking with digital film became a substantial part of marketing the film before and after its release, with the newly developed apparatus for capturing the effect revealed and discussed. Bullet-time seemed to find its way into everything for a few years and we still feel its effects in the dance videos that we're discussing here, and in a huge variety of other visual media –it's become part of the landscape of our expectation. In the 'PUMPED UP KICKS|DUBSTEP' clip, at 2:25–2:33 minutes, the sense of weightlessness Nonstop evokes seems oddly resonant with Neo's movements somehow, but is perhaps more closely related simply to slow motion; the reference seems explicit, however, in the 'TAKE ME AWAY | DUBSTEP' clip. At 0:53–1:08 minutes, Nonstop's movements quote Neo's slowed-down move to the floor before mimicking a more traditional mime form to get to his feet and then return to his knees again. And again at 1:24–1:39 minutes, *The Matrix*'s slowed-down time seems to be in play – we're a camera move away from being able to observe Nonstop's dancing at Neo's pace with *The Matrix* viewer's power.

Of course, *The Matrix* isn't the only thing influencing the movements in these performances, but bullet-time has become part of a whole culture of visual cues that form a collective grammar of bodies observed in motion, visual cues that act as prompts for movement and frames of conception for interpreting what's on display. Like the mirror in the dance studio, bullet-time functions as a reflexive technology, something which causes us to reconceptualize our potential.

A New Aesthetic

The suite of elements that make up popular visual grammar are a broader concern than reflexive relationships; reflexive interactions are those meetings with technology that prompt us to reconsider the potentials of our embodiment via grammatical manipulations, but visual grammar is also constantly iterating around new ways of looking at the world.

There has been an attempt over the last three years or so to try and describe a 'New Aesthetic' in the visual arts, one centred around digitization and the intrusion of computing into visible real-world experience. An image search of the term will start to give you an idea of what we're discussing. Note the use of pixels, glitches, wireframes and computer-readable rather than human-readable symbology – this is a new category of visual artefacts that we often either can't read, as humans, or that we shouldn't see.

The New Aesthetic isn't, or isn't always, a thing, something you can simply point to or pick out, but I suspect that it's real, or at least partially definable. It is perhaps best described as the act of curating objects with bonds of affinity arrayed around the intrusion of computation and digitization into our everyday lives, an act of de-familiarization, of collecting what increasingly surrounds us in plain sight. Its artefacts are often unreadable, often ignored and yet impact upon the seen, heard and felt world that we surround ourselves with (we might even see it as the seeking out of rules for a new visual grammar). In the words of James Bridle, the writer and artist who first outlined it, the New Aesthetic 'is a series of artifacts of the heterogeneous network, which recognizes differences, the gaps in our overlapping but distant realities' (n.d.).

The New Aesthetic, at least as something described, has a couple of start dates. Bridle's blog post on 6 May 2011 pointed towards a new Tumblr[15] that would, initially at least, collect images of current technologies with a sense of 'wonder' (see note 15), technologies that seemed to point to a new way of conceiving of the future. This sense of newness, however, doesn't cover the intense nostalgia for early, 8- and 16-bit computer graphics that have since come to be one of the defining aspects of objects identified as being part of the New Aesthetic, and even Bridle's early postings attest to the illusion of any simple coherence arrayed around wondrous objects and future gazing. This date then, 6 May 2011, represents the start of the Bridle's (art/research) project called 'the New Aesthetic', but its birth as a debatable cultural concern is more reasonably placed at either the 12 March 2012 South by Southwest (SXSW) conference panel organized by Bridle,[16] or the science fiction writer Bruce Sterling's 2 April 2012 write-up of the event for *Wired* magazine which marked the start of the serious and sustained attention on the discussion that Bridle had begun.

But the New Aesthetic is, of course, also older than its formal start dates, impossible to pin down. As Paul Cézanne's series of paintings of Mont Sainte-Victoire represented a new way of depicting subjective experience (increasingly trying to capture the movements of his head as he painted, his intimate personal experience of colour in that moment, even his saccades, the almost imperceptible side-to-side dance of his eyes as he built his seen-world), in much the same way, so we see a New Aesthetic emerging which brings out the almost imperceptible, as well as the explicit influences of our cyberculture, a New Aesthetic which I would argue has also made itself felt in the contemporary dance performances described above. And, again, this isn't new; emergent grammars have always influenced the production and reception of new visual arts. Marcel Duchamp's 'Nude Descending Staircase', for instance, can be hard to process until you've seen examples of stroboscopic photography. Image searches of both the painting and 'stroboscopic photography' break down the mystery of Duchamp's work and reveals the effect as an analogue precursor to the sequence photography discussed above.[17] Knowledge of the effect quite clearly offers a framing context for the image, moving it, for many viewers, into the realm of intelligibility.

Each of the artworks and performances discussed here present another way of looking at the world, another way of depicting motion, of action over time. That this kind of play can be revelatory was shown by Eadweard

Muybridge's 'Sallie Gardner at a Gallop', the early photographic experiment which first captured the fact that all four of a horse's hooves leave the ground during steps at full gallop. Such instances in the history of art and science (and their repeated crossover) offer a real awareness that our visual experience, our sensory experience in general, that bodily stuff that's meant to avoid influence, is just as shaped by our past as our thoughts and speech and sense of self. All of these seemingly new ways of looking at the world are often the coming to consciousness of the ways in which we've already been warily looking, but that have now become heightened before being normalized. These sorts of things are at their most powerful, I would argue, when they function as reflexive technologies, changing the grammars by which we conceive of ourselves, our surroundings and our resultant potentials for action. The New Aesthetic, as a project, starts to capture the newly reflexive effects of our digital technologies.

Lasting experience and expertise with these things changes the ways in which we look and act; our lines of potential are written differently in the world around us. Anyone who sketches, or takes photographs, or skateboards will be familiar with this, those times when you can wander around in an environment seeing it in terms of forms and lines, geometric planes and shades and vectors, paths and routes, rather than lived spaces and typical uses. That word 'grammar' seems to describe these features of acting too, to describe those rules that we know, but don't know that we know, that condition how we receive inputs and help to organize and emphasize our expression. Reflexive relationships with technologies are what allow us to newly say: 'I can make *that*, I can do *this*, I can process *this*, I can go *there*.' The same world, but not the same, a world conceived of in a productive reduction alongside and through those artefacts that co-determine our encounters and facilitate and multiply our actions, endlessly.

A Response

Mapping the Prenoetic Dynamics of Performance

Shaun Gallagher

In the study of embodied and situated cognition one important issue concerns identifying precisely what aspects of embodiment and situation affect our lives – both in terms of how we perceive the world and deal with it cognitively and in terms of our social relations. Once we escape internalist thinking that would reduce everything of importance to neural processing, or in more traditional, Cartesian terms, the mind – once we start to acknowledge that the whole body as it engages with its environment shapes our cognitive and intersubjective life – then we need to know how to map out the dynamics that characterize such processes. There has been a lot of research on sensory-motor processes, including kinaesthesis and proprioception, as well as the role of sensory-motor contingencies in perception (e.g. Gallagher 2005; Noë 2004). Following the lines of distributed cognition and extended mind approaches, there has also been exciting research focused on the role of artefacts (e.g. Malafouris 2010), tools and technologies (e.g. Clark 2008; Hutchins 2014). Once we notice that such things can not only extend our cognitive reach, but also constrain it, we can also ask about the constraints imposed by our own scientific tools and methods. We can start to see, for example, that putting an individual into an fMRI scanner and showing her still images of faces or even action videos will not give us an adequate account of social cognition, which for the most part involves interactions among people in the world and not just neuronal interactions inside a machine.

One question that I've been interested in concerns not only what the body, and the things, and tools and technologies do for us, for our cognitive life and our intersubjective relations, but also what social and cultural practices and larger-scale institutions contribute to such processes (Gallagher 2013). This extends the notion of extended cognition and acknowledges that the distributed aspects of cognition sometimes are not equally distributed but tend to clump or coagulate around established structures and norms. This can be for the good and can extend our knowledge. Science itself is one such social practice and institution (Slaby and Gallagher 2014). It can also introduce distortions and narrow down possibilities – we can find many examples of this in architectural design, educational organization, social and political structures and so on. Sometimes the really powerful practices and institutions, however, are the ones that we don't notice, or ones that are so intrinsic to our lives that they remain invisible.

In previous work I've used the term 'prenoetic' to signify certain bodily processes that we are not aware of, but that shape the way that we perceive and move around the world. Body-schematic processes, for example, work this way; we are for the most part unaware of such motor control processes, but they influence the kinds of things that we are conscious of, and the kinds of actions that are possible for us. Likewise, a large range of bodily affects remain prenoetic (unknown to us as they occur, but shape our perception of the world and of others), and in some cases we find ourselves making decisions (exercising our *noetic* or thinking abilities) under the influence of such affects. For example, the bodily affects of hunger can bias judicial decisions that ought to be governed purely by legal reasoning (Danziger, Levav and Avnaim-Pesso 2011). This notion of the prenoetic can also be extended to environmental and institutional factors. The settings and the norms and the rules that have evolved over time to guide legal processes shape the way that we make judicial decisions, and many times we are not fully aware of their effects. The proceedings are conducted in *this* way, because that's the way it is done, and has 'always' been done; and yet doing it *this* way may in fact lead the judge to make life-altering decisions just before lunch when she is most hungry.

Research into prenoetic effects has to dig deeper into particular examples to learn how these things work. The chapters by Hayler, McCarroll and Tribble provide extremely productive analyses that do dig deeply into bodily and social practices and the extended effects of technology to reveal the surprising ways that they change our perception, our way of thinking, and also loop around to recursively change our bodily and social practices.

Matt Hayler, building on insights found in Don Ihde's post-phenomenological analyses, considers the effects of media on our perceptions and actions. Just as artefacts and the various machines and technologies that we use can be incorporated into our body schemas, or can filter what we see and how we interpret the world, and can thereby alter our possibilities for action, much of this can happen prenoetically – in the background and without our noticing it. In the same way that as we speak and communicate with others, we depend on grammar without thinking about it – indeed, we learn to speak without any inkling of grammatical rules and only come to learn there are such things in 'grammar' school – so our use of things and technologies involves implicit (I would say, prenoetic) grammars that work in a reflexive way involving looping effects (Hacking 1995) that can either expand our affordances for action, or constrict them.

Hayler points to the example of photography. He nicely shows how a certain grammar comes along with techniques, such as tilt shift photography, in a way that manipulates our perspectives. Photographic techniques can change the way that we see things; in some cases it can make visible what had previously been invisible. People had been watching the movements of horses for centuries, but they were unable to observe accurately the galloping gait because of its high speed. Even experts had no clue to what the gallop was truly like. In the early nineteenth century, for example, Théodore Géricault, the painter and jockey who had gained prominence with his famous 1819 painting of *The Raft of Medusa*, went to the horse races in search of new imagery, and in 1821 he painted *The Epsom Derby*, now in the Louvre in Paris. This painting reveals how Géricault, an expert in both painting and matters equestrian, an expert, therefore, in both visual representation and in what was being represented, saw the horses' galloping gait. At full speed the horses simultaneously stretch their front legs out ahead and their back ones to the rear. Most equine experts also 'saw' the horse's legs in this position during a full gallop. Speed photography changes this perception. As Hayler notes, the exact motion of a galloping horse's legs would only be revealed when Eadweard Muybridge's serial photographs were published. In 1872, he produced a sequence of photographs of a horse at full gallop and then displayed these images in quick succession, creating the effect of a film. This visual sequential depiction of the motion showed unambiguously that Géricault's earlier assumption was wrong; a horse's front and rear legs never simultaneously extend off the ground away from its body during a gallop. The crucial point, however, is that subsequently we all began to 'see' horses differently. Anyone who views Muybridge's film *Horse in Motion* will thereafter also see every live horse's galloping gait differently. The film influences the act of seeing (Barber 2012). As Jörg Trembler notes, we do not have the feeling of having understood something, but rather the feeling of *seeing* something directly, when before we couldn't see it at all (Gallagher et al. 2015).

Hayler shows that the use of mirrors, photography, cinema and video cameras continue to have reflexive effects on how we see and understand things, and even on how we move, specifically as it relates to dance. He shows how prior to its time a specific style of dance 'couldn't even have been *conceived of* in terms of the effects that it deploys because the motivation for those effects emerge from the cultural baggage of over 40 years of digitally tampered-with visual imagery and 200 years of photographic experiments' (247). The implicit grammar that allows us to see what we see and know what we know gets built out of that overexposure, forming 'rules that we know, but don't know that we know' (257). Such an evolved grammar forms our expectations, shows us different possibilities and, in effect, creates new affordances for action.

What Hayler says of media and technology, Sarah McCarroll shows to be the case with something even closer to us – our manner of dress. Beyond the proverb, she shows that clothes really do make men and women; clothes impact most immediately how we move, and then how we act and what roles we can play, helping to construct specific social structures that again loop around to reinforce the customs and costumes that we don. McCarroll builds on the distinction between body image and body schema (Gallagher 2005) and shows that the clothes that we wear are not simply a matter of dressing up our body images, but can actually take hold of our body-schematic processes and, within specific social settings, operate as a means of colonizing our actions. Clothes can impose a specific behavioural pattern (McCarroll refers to it as a 'body map') on our actions by defining (delimiting) movement. McCarroll views 'the acquisition of a body map as a largely invisible process as strictures of polite behavior, structures of clothing, and saturation of visual imagery act upon the consciously adopted habits of dress and behavior related to body image, and permeate the pre-conscious body schema' (146). Here she demonstrates her point with the example of the corset which acted as something of a straightjacket on the bodies of Victorian women and defined their role in society. Her evidence is found in J. M. Barrie's play *The Admirable Crichton* (1902).

Clothes, and more generally, fashions, are like an institution that we wear. They can impose rigid limitations on movement and on daily practices and seriously shape our social customs. In Victorian London, the Lazenby daughters in Barrie's play require intensive attention from their individual maids simply in order to dress. Dresses are buttoned from the back; corsets prevent the women from bending to tie their own shoes, which required stylish and complex lacing up. The corset and everything that goes with it – all the invisible and unmentionable garments – rob them of free movement and prevent them from engaging in certain types of action. The daughters are in effect dressed to be dependent and helpless and are pushed into a very restrictive, corseted social structure.

We may think that today we've been liberated from these types of clothes – although, of course this is not the case in all cultures, and particularly there continue to exist strictures on women's dress in the name of modesty, decency, God or business acumen. We might point to some of those obvious examples and miss the fact that even changing and liberating fashions nonetheless continue to be imposed, since a woman is still expected to dress like a woman and man like a man – especially in specific settings. Whether it's tight jeans and high heels, or the very loose clothing that seems always to be in danger of falling off, or the traditional business suit – these are standards that shape expectations and define the norm. Clothes continue to be institutions. The larger point here is that institutions continue to be institutions, and we often find ourselves in good ones, like loose and comfortable clothes that permit a lot of free movement, or in bad ones that tie us up in tight and constrictive practices that discourage innovative actions. Institutions, like clothes, also define an affordance space – a set of possible actions across a range of physical and social settings (Brincker 2014). Clothes are more than metaphors in this regard since they have real physical and social effects and can actually support the norms of institutions.

In a similar fashion, Evelyn Tribble shows us that you can dress up an expert, but you can't take him or her just anywhere – expertise, to the extent that it's a form of distributed cognition, will depend specifically on features of the physical and social environment. A surgeon who is an expert in the hospital's well-equipped and well-lit operating room is less of an expert if the surgery has to be performed in a jungle or in a jet plane. Expertise frequently requires a well-defined niche. This applies to acting too, on stage or on screen. Tribble cites John Sutton's work in this area, and focuses on the idea that expertise is a 'mesh', a kind of vertical integration between the low-order flow of embodied coping emphasized by Dreyfus, and higher-order, more reflective cognitive aspects. For the actor this mesh corresponds to a combination of 'preparatory thinking as she readies herself for the role, and in-performance thinking, which, in an ideal situation, is "aligned" with the actor's action' (138).

We can also emphasize something like a horizontal mesh that incorporates parts of the physical and/or social environment in both preparation and performance. For example, Sutton has also shown experimentally that how good one's episodic memory is may depend on one's partner, so that two people who have been long-term partners can remember more through their conversations than when the two are asked to recount memories without interacting (Sutton et al. 2010). Two in conversation adds up to more than two in their individual performances.

My daughter Julia is a professional actress and, in that context, I was always impressed by her ability to memorize her lines. She explained that at least for some actors blocking, that is, the working out of position and movement of the performers on the stage, helps to support memory. When one is in a particular position at a particular time in the play, interacting with another performer, the lines are also there – lines and gestures and postures and movements are elicited by the details of the scene. I once had the opportunity to ask the actor Richard Gere about this, and he explained that although there is also blocking, made more complex with the addition of camera position, in shooting a film, remembering lines is further assisted by good writing. The quality of the narrative and the flow of the lines also mesh with the flow of the actor's embodied ability and their understanding of the character.

Acting clearly involves a distributed expertise. I would add that actors have to be experts in being novices. Actors may have certain techniques that allow them to perform with excellence. But for each character they have to begin from scratch, or more precisely, with whatever knowledge they have. Then they have to unlearn much of that as they take on the new character and the new performance; or as Tribble puts it, 'the actor ... must banish certain forms of thought and harness others' (138). As she indicates, it involves the meshing or harnessing of Richardson's four horses – which also suggests that only by being an actor and being inside the process does one discover how it really works. Acting involves a transformed perception that cannot be attained from the outside or in an unassisted viewing. Good actors make a difficult process look easy. To get the complete picture means to go beyond a snapshot analysis and into the meshed dynamics of performance that might only be caught by a Muybridge-style revelation.

The chapters by Hayler, McCarroll and Tribble start to give us this kind of insight into performance. They show that perception, cognition and performance are not the accomplishments of narrow processes taking place in Cartesian minds. Not only expert know-how and artistic expression, but also everyday movement, action and communication – these processes are both enabled and constrained, sometimes pushed to the level of superior performance by technological and social scaffolds, sometimes distorted by poorly designed institutions, and sometimes defeated by overly constraining norms that limit affordance spaces. Such elements have prenoetic effects that in unsuspecting and surprising ways can transform our perception, our way of thinking and also loop around to recursively change our bodily and social practices.

After Words

These 'After words' are 'origin stories', interviews and written pieces describing the ways the work of some practitioners has been affected by the cognitive sciences. The co-editors' stories are echoed both in our interview with John Emigh, emeritus professor at Brown and an early adapter of scientific research, and in the essay written by Deb Margolin, playwright, performer and original member of Split Britches. We spoke to professional artists about their work, and how they think about the creation and reception process. We hear in these voices the artistic instinct and the urgency of the practical realities of the rehearsal room.

Origin stories

Rhonda Blair: In the latish 1990s I was drawn to cognitive science because it provided a holistic way of understanding experience and being, particularly in relationship to my theatre practice and research. What I knew intuitively from my intellectual and artistic work couldn't be accounted for by the compartmentalized approaches embraced by some of my gifted colleagues in acting, who had an anti-intellectual and short-sighted fear that theory couldn't coexist with creativity or feeling, while some of my very smart colleagues in feminist theory could talk brilliantly about a theorized Body with a capital 'B' while marginalizing feelings and the experiential, reluctant to engage the full range of the body's complexity and messiness in the acting studio. (Interestingly, a space is now being made for this in some affect theory.) Like John Emigh, I love the messiness and complexity allowed by engaging the cognitive science.

Amy Cook: In the latish 1990s I was an assistant director on Amy Freed's *Freedomland* at Playwrights Horizons in New York City and she and I would talk about laughter, how it was evoked, how it could be blocked, and I began to wonder if there was neuroscience research on laughter and perception. I then directed a found-text performance using Ranier Maria Rilke's *Duino Elegies* and I was struck by how audiences responded differently (and responded in unison) to the language of poetry onstage. I took a neuroscience course at Hunter designed to give students an overview to different approaches and I felt sure that integrating new research from the brain sciences could inform our understanding of bodies speaking language onstage. Nothing I had read at that point sufficiently explained the power of theatre to make up my mind and I needed additional tools to find new answers.

An origin story

An Interview with John Emigh. Emigh, an emeritus professor at Brown University, was one of the earliest in the arts and humanities to think about the implications for his research in theatre – in particular, the masked performance of Bali – of research within neuroscience.

Cook: You were one of the first people that I knew of who integrated cognitive science into questions of performance, in the 1980s I believe. That's how I think of you, but that may not be how you think of yourself. To what end have you turned to cognitive science and why do you think you did so?
I think of myself more as a practitioner, but I'm probably known more as a scholar or an expert on Asian theatre. Things in cognitive science have fed back into my practice because they've been inspirational or because they've helped me sort out different ways of thinking and offered different vantage points – basic questions about the relationship of self to character, or about the way narrative leads towards or is conditioned by or acts with present-day circumstances. Elements in performance often go on automatic pilot or run by habit, and sometimes it's interesting to interrogate that with a different vocabulary and set of working conditions like cognitive science to see if what may seem exotic from the appearance of things in fact has within it principles that aren't really exotic at all. Cognitive science can help us rethink the way in which things live on stage and in the audience's mind.

For me cognitive science is just another way of trying to break up and question assumptions. It might reinforce certain ideas that were already intuitively there, and it might make me wonder about some other things that intuitively I feel don't work or that might work better another way, so occasionally I'll make references to cognitive science research in teaching acting or directing, but not really that often. I haven't used it as a recipe.

***

Cook: How did this begin?
When I was doing my MFA in Directing at Tulane in the 1960s, Richard Schechner convinced me to start doing a PhD at the same time, so I was

moving back and forth between theory and practice. I don't even remember which one of us initiated it; it just seemed natural and I had to take some theoretical and historical classes anyway and I was interested in the classes, and I had a dual major in English and Theatre at Amherst. It was, if not follow my bliss, at least follow my curiosity. So when Richard was getting interested in a kind of anthropological approach to theatre and in attempting to push the borders of Western theatrical knowledge and in incorporating theory and practice, I was right there for him. I was ready for it. That was easy, that was a 'Why not? Take a year or two more'.

***

Cook: The understanding I had of the brain from college did not welcome integration with theatre at that point. But then the research started changing. When did theatre and neuroscience come together for you?
I was chair of the theatre department at Brown in 1991–2, and at the same time I was teaching the history of dramatic theory. I had a couple of students come in who wanted to concentrate in theatre, but who were also interested in concentrating in this thing called neuroscience. They seemed very ashamed and apologetic, or at least they seemed to be asking, 'What's wrong with me that I'm interested in these two things that have nothing to do with each other? I don't know what to do with my life' or 'What am I supposed to major in?' and 'What's the aberration here?' So I asked them, 'Well, what do these things have in common? What is it that attracts you to them?' Theatre is all about attraction and attention, about what is compelling; it's about how we parse meaning, how we communicate meaning, how the living body is somehow emblematic of a mode of thought, and moves back into thought. If neuroscience had nothing to teach us about these things, then what the hell were they thinking about? Now certainly, for example, questions of empathy come up in theatre: How does one look at the actions of another and make connections with your own life? There's nothing new about this in theatre, but it was odd that there seemed to be such a disconnect between the two different disciplines. I got to thinking about Freud's old work with jokes and Arthur Koestler's related work on creativity, and I thought there might be some stuff there to explore. This was about the time, of course, when neuroscientists themselves could no longer avoid the large questions. Damasio was coming up with his body-oriented brain modelling, and Daniel Dennett was absorbing the work of cognitive and neural science into his philosophical writings.

One of the fascinating things to me about the work in cognitive science in the 1990s was that, on the one hand, Damasio was reinventing Aristotle's *ethos*, and the way in which one forms an *ethos*, an identity, in relationship to repeated actions and usages of the mind and, at the same time, Daniel Dennett was working through the very post-modern conception of the haphazard enactment of self from the multiple drafts the mind makes on the fly, and that, almost by chance, we do 'this', instead of 'that'; Damasio and Dennett were looking at the exact same evidence and they're diametrically opposed, and I loved that. In performance studies at around that same time some scholars were attracted to cognitive science, while it seemed that others didn't like its messiness. And I loved that messiness. I loved the fact that people didn't know quite what to do with these connections yet, and they were coming up with diametrically opposed answers.

Cook: What do you want the cognitive sciences to explain? Do you think they can address questions actors have?
On an intuitive level, actors of course are very alert to what audiences are responding to or not responding to, for very practical reasons. They know a great deal, and it would be surprising if cognitive science came up with things that were alien to that experience. It can be helpful to understand what is happening through a different linguistic and conceptual prism, as long as it facilitates the work and doesn't make it overly self-conscious.

***

Cook: Can you talk about how a theatre project you did was affected or informed by the cognitive science?
I first got attracted to theatre, really, because I was just trying to learn Spanish. I was reading the plays of Federico García Lorca; I got fascinated by his use of very complex sets of images that are interwoven with linguistic practices of southern Spain, and a kind of wonderful purchase on metaphor, and a kind of poetic recreation of theatrical possibilities through that metaphor. I went to Spain in the middle of college, back in 1961, partly because I didn't know what I was doing in college and partly because I wanted to translate Lorca. In Lorca's play, *Así que pasen cinco años* (As Five Years Pass), a wedding dress comes to life, a larger-than-life wedding dress that, in this extraordinary poetic script, taunts the failures of the central figure to live up to the responsibilities of Spanish manhood and have a child. Of course, Lorca was certainly dealing with being gay, and maybe beyond that, issues of sexual identity and gender identification. It's a kind of nightmare vision of his going to America and coming back to Spain and not having children. Here's where cognitive sciences help. When you're dealing with something that is familiar, metaphorically, it's very simple, it just recedes into normal communication and you process that with the verbal processing of the left

brain. When metaphor becomes complicated (this is related to Lakoff and Johnson's work), you need to engage the right brain, because you need to get visual images, and you need to reorient what's happening in the organization of these images in the brain. You have to parse it differently. As you do that, you maybe take away some of the activity that the left brain is available for.

The keys to doing the play, which I co-directed with choreographer Michelle Bach-Coulibaly in 2003, were original music and the wonderful contributions from the costumer. In sections where it became very rich in dance and music, we built in some redundancy and slowed down the verbal process. And where the verbal processing became very rich, it was necessary to bring relative stasis to the movement aspect. Sometimes when I see colleagues working more abstractly and incorporating more dance into spoken text work, I find myself, as an audience member, unable to maintain proper attention, because my attention is too divided. It puts certain a stress on the brain. The audience is trying to create this cognitive sense of what's going on, and to orient it in relation to their own lives; they're shifting their attention back and forth between their musings and what the words are actually saying and the metaphoric reach of the words and the appreciation of the sensory overload that's going on and the flow of music and dance and imagery and Spanishness and whatever else is happening. A huge giant person with a dress billowing out, standing on another actor's shoulder, becomes a kind of terribly threatening figure of a bridal dress that's never worn. On the one hand I love all that richness, but on the other hand, there comes a point where the audience just gets confused and shuts down, if the richness is calling on different abilities to parse things out in too many ways at the same time. So, as a director, it became very, very useful for me to understand cognitively what the problem was, which I may have had an intuitive sense of anyway; not having this mode of problem-solving is probably one of the things that hadn't allowed me to do this play before.

***

Cook: Talk about your work with neuroscience and masked performance.
The conversation with the confused freshmen wanting to concentrate in both theatre arts and neuroscience spurred me to take advantage of a grant opportunity to offer, outside of my course load, and exploratory course on possible connections between theatre and neuroscience. Three students and I had weekly conversations with various professors from cognitive science, neuroscience (the two fields were then still separate and suspicious of each other), engineering, and linguistics. It was fascinating to me and to the students, and to my surprise professors from these disparate fields were

also eager to learn how their research and disciplinary concerns related to theatrical theory and practice. I taught this course about every third year until my retirement from regular teaching in 2009. I always taught this class with the aid of advanced students from the cognitive sciences, and while I established broad categories of inquiry and suggested core readings in each category, the actual reading lists and class discussions were always led by student teams. At first I had to take great care to have about one half of the class from the arts and the other half from the sciences, but in more recent years, many of the students were well experienced in both realms. Clearly, the 'Two Cultures' or art and science are coming together – at least in the academy. In the process of teaching this course, I got very interested in Penfield's drawings of the homunculus and took up a clue in David Napier's work, and extended it – especially, but not exclusively, in relation to Balinese mask performances I'd be working with and performing in for several decades.

***

Cook: In your 2011 essay for *JDTC*, you explore demon and apotropaic masks/mask-life faces to find a particular facial expression that seems almost universally to represent the demonic and you suggest that perhaps it is the particular blend of human and bestial features that makes the masks so evocative. Maybe the features are made larger because 'studies of visual tracking indicate that the eye focuses obsessively on the mouth, nose, and eyes to "read" a human face' (Emigh 2011). Your attention remains on the cultures, while turning to the sciences referentially. How has cognitive science informed your understanding of culture?
Was there a way that cognitive science could interface with notions of culture? I didn't want to destroy the usefulness of culture as a concept. I wanted to pare it down to its limits, because 'culture' is this kind of heuristic word, it's a bootstrap word – there is no 'there' there. It's something that we make up in order to talk about things. But maybe there is a 'there'. Maybe the 'there', if it's anywhere, is in the human brain. It's in the brains of people who live in association with each other, and who share stories and bits of moral precepts, and ideas about things, some of which they accept, some of which they reject, some of which they find conditional and some of which they contest among themselves. So culture, both as it presents itself in interpersonal relations and as its elements are manifest in the behaviour of individuals, exists as an integrated – but certainly not static and often contentious system whose material existence is in the minds of those involved. It's not unlike the contentious society of the mind that Marvin Minsky writes about and that Edelman wonderfully extends, describing how we hold categories in our

head and contest them and shift them around all the time. Is there a way of thinking of this as a cognitive network in which we operate, where we are quite free to disagree with each other and even with ourselves at times, as we all do? Is that the physical place of culture? This view of it has the advantage of being seen as always in flux and contestation, and which also allows for individual difference.

The model of culture used in the past didn't really do very well, so there were problems: how do you allow for individual difference and how do you allow for dynamic change within a culture, and therefore escape the kind of fixity of the anthropological present and the insistence that people (not, God forbid, in our own institutions or in our own places, but in those other 'exotic' places) all behave the same. This has been one of the most useful ways that cognitive science has helped me understand the way in which theatrical practice relates to societies – if you're working in a place like Bali, then you're going to be confronted with things that just are beyond your ken.

***

Cook: As a practitioner as well as scholar of Balinese performance, do you think you experience or sense the science when you are working in a different way than if you simply studied the science? Can you talk about how cognitive science is related to your work in Balinese performance?

I spent a year in Bali and learnt Balinese dance, and that challenged me to think about what constitutes the self and what's the plasticity of self that allows you to wear different masks: What's the work in trying to wear a new mask, and to adapt to the mask or bring the mask into some sort of living association of self and your own life and physical possibilities? For example, the Balinese are very clear that it's a big mistake to think of what a Topeng performer does as possession or a trance. They certainly talk about having *Taksu*, which is like 'having presence', or being in 'the zone' or in a state of 'flow', and feeling inspired and going beyond what one could logically plan as possible in a performance. But they're quite clear that that's different from a loss of consciousness. In fact, it's a hyper-consciousness; this is different from the loss of consciousness that happens in trance performance, which also takes place in Bali. Very rarely do Balinese performers do both – that's quite rare. Descriptions related to this aspect of heightened consciousness go as far back as Zeami (the late-fourteenth-to early-fifteenth-century playwright and actor) who talks about the hyper-alert state of things in performance – of being in the flow of activity and yet being an outside admirer or guide to that activity at the same time. One of my teachers said very similar things about Balinese Topang performers when they are being most effective. The senses are hyper-alert to everything that's going on: everything comes into the body and performance through the learnt craft, through the kinaesthetic traces, through everything that the body knows, but it becomes extraordinarily alive in that moment and the performer is hyper-aware of everything happening.

***

Cook: How is this connected to Western acting for you?
That whole aspect of understanding consciousness has fascinated me, in terms of extending what an actor can do in relationship to character. How do you get beyond the limits of your performed persona in day-to-day life? The huge limitation of the way in which Stanislavski's teaching was taken by Strasberg and brought into the American method was to require the actor always to justify what a character does in relationship to the actor's personal lived experience; this puts extraordinary limitations on what a character can do. As Michael Chekhov himself complained to Stanislavski, you have to allow space for the imagination. The self is larger, it's more whole than our 'daily' sense of self. In Indian philosophy, the microcosm contains the macrocosm, each individual body contains the richness of the whole universe. It isn't always expressed, it isn't always made manifest, it isn't even often realized by the person possessing that body, but there's a richness there. Once you recognize that there are angels and demons within each of us, there's a whole range of human possibilities in us that we don't express in our everyday life. How do you get an actor to be comfortable in pushing and extending that range, and feeling a sense of it all happening in their own body? It's not that they're feigning being something they're not, but how do you claim that potentiality?

In his 'Essay to an Actress' Freud actually talks about this: How do you get the untapped possibilities of the performed self in everyday life into acting? Although he wasn't sure how this took place, Freud's understanding was that somehow what an actor does was a controlled state of dissociation. You have to dissociate yourself from the social self that you've adopted as a persona, whatever aspects of this are dictated by nature and by nurture, in order to extend the possibilities of the living self in performance. You have to find the meaning between the living body and the mask, or the actor and Hamlet's words, or whatever it is that you're trying to do. How do you make some sort of amalgam – not just a conceptual blend, but an actual living blend, an embodied blend, of the actor's body and the words or actions or character traits. In order to do that, I think you need to have a controlled active dissociation. If I think about the various acting exercises that have been passed down to me and a few of which I've come up with myself, they really can be thought of as ways to encourage and control that active dissociation (whether, for example, it's Chekhov's work with imaginary centres or Meisner's work on hyper-attention to the given circumstances).

One can talk about this in lots of different cultures and practices, but in some ways, I often get the feeling the exercises are like Dumbo's feather. They're simply a way of giving licence to the imagination and for the body to be other than the constrained, everyday body, by setting in motion some sort of other task. Dumbo didn't really need the feather to fly, but, yes, he did need the feather to fly because it gave him permission to not be grounded. So in some ways, what you're doing as an acting teacher is handing out a bunch of Dumbo's feathers, in the hopes that one of them will give the actor, the student actor, the freedom to go beyond the bounds of his 'person'. And then you add craft to that as you go along.

***

Cook: Any final thoughts?
We have thousands of years of being in a workshop in the nature of identity, which is what makes theatre so challenging to orthodoxies or to churches. We are living demonstrations of the plasticity of identity, and that's terribly challenging to any kind of orthodoxy. Whatever approach I'm using, it's not so much a matter to me of finding the answers, it's a matter of keeping a dynamic inquiry going. Which doesn't mean that you don't ask the big questions, or that you don't venture into grappling with the big questions.

I think it goes back to Csikszentmihalyi's flow state. I think he's grappling with the question of what is so satisfying about doing something when we feel that sense of 'flow', when we simply (or quite complexly) are what we're doing. There's not all this chatter about 'should we or shouldn't we be doing it, or might we be doing something else?' There's just the complete immersion in activity, what in sports is called 'being in the zone'. One of the reasons why people, certainly in college, take up dance or theatre is that, after a lot of hard work and rehearsal, you get to be fully something. You live the life of not an abstract Hamlet or Horatio, but Hamlet or Horatio as they exist from moment to moment in your own body, all, of course, in accordance with directorial choices. Words come to embodied life in some sort of meeting ground with your own physical and psychological self, and there's something wonderful about being able to do that. Performances, though, are fashioned and enacted through ongoing cycles of analysis and synthesis, engaging both brain and body in a recurring process. In the analytical phases of these cycles, perspectives drawn from the cognitive neurosciences can help to clarify the issues and goals at hand; such perspectives will no doubt eventually help us to understand the ways in which the body-minded brain engages in the synthetic phase of enactment. These perspectives are already of great value in understanding how audience members – individually and collectively – parse, identify with, contest and process the words and actions performed.

An origin story

Deb Margolin:
Original member of Split Britches, Deb Margolin is a playwright and performance artist. She is the author of eight solo performance pieces that she has toured throughout the United States; she is also the recipient of the 1999–2000 Off-Broadway Theater Award (also called Obie Award) for Sustained Excellence of Performance and the Kesselring Playwriting Award for her play *Three Seconds in the Key* in 2005. She was awarded the 2005 Richard H. Brodhead Prize for Teaching Excellence at Yale University, and had the honour of accepting the 2015 Helen Merrill Distinguished Playwright award. A compilation of Deb's performance pieces and plays, entitled *Of All The Nerve: Deb Margolin SOLO*, was published in 1999 by Cassell/Continuum Press, and a video recording of the performance is also available through Metropolis Media Productions. Deb is currently an associate professor (adj.) in Yale University's undergraduate Theater Studies Program, and a proud member of New Dramatists.

We asked her to tell us about how her artistic work has been informed by science:
As an English major at New York University in the 1970s, I spent three years reading books and poetry and writing about them. During this time I thought I was Henry Miller, because I considered myself a writer, and specifically a writer about and from the body, and there were absolutely no women I'd been told or taught about who did that. (My discovery that I was wrong was delicious and came later.) By senior year, I realized I knew nothing about the material world, the world of cells and organic activity, the nature and shape of bodies both celestial and human, and so I dropped all literature classes and studied hard sciences: lab biology, chemistry, psychophysics. I studied the Krebs cycle, the lunar cycle, the menstrual cycle; I cut open a goat's eyeball and felt awe at the nacre of the retina; I injected Jack Daniel's whiskey into an ecosystem of amoebas and euglenas and parameciums and when I came back they had quadrupled in population! (Ah, Jack!)

All of these were richly illuminating, but it was my psychophysics class, a course studying the relationship between stimuli and perception, that really changed my philosophy as a thinker. The professor was kind of a grouchy and slightly scary man who didn't suffer fools gladly, and most of us were fools in his opinion. Nevertheless, I really liked him. He had this gravelly, gruff sense of humour, and a good sense of the absurd, and I laughed quite a bit under his caustic auspice. During one particular segment of our study, something revelatory was said by him. We were studying the relationship between light and the eye and the brain, and the many, many miracles that transpire to create vision, sight. We had to discuss the divine mystery of light as both a wave and a particle; to discuss colour and metameric pairs; then we moved into a discussion of the eye's reception of light; of rods and cones, the perception of colour, the movement from these neurons of the retina to the brain of an image, upside down, the brain's righting of that image. A million miracles at breakneck speed, one after the other! Then, as we discussed the blind spot on the retina, the part that is devoid of perceptual neurons, a fairly significant piece of real estate on this sensitive surface, one student found the courage to ask an important question. (Asking a question of this teacher risked ridicule and worse.) He asked the professor why, if there were these two huge blind spots in the eyes, we don't have a corresponding hole in our perceptual field. Why isn't there a hole in our vision? And the teacher responded that

> *we don't notice this absence because we weren't wired to pick anything up there in the first place!*

This is really revelatory! It filled me with glee and humility! It meant that I was an absolute servant of my perceptual apparatus; that I perceived of the world only what I was *wired* to perceive! It meant that there were worlds that most probably exist, sounds, colours, moans from trees, wheezing of stars, that existed beyond my perceptual ken! It meant that I was a sex slave to my own body! It freed me to my imagination! The clarity of this limitation opened up the world!

Years later when, after calling Mark Russell, then artistic director of PS122, every three months for five years, he finally agreed to give me a performance slot at this exciting downtown New York theatre, I brought a full-length solo show, which included the nerve cell monologue in this collection. I wanted to call the entire show

Of All, The Nerve

because this revelation that I was only my body, a realization that opens doors for actors, for poets, for arguments about the nature of God; this idea that I cannot know more than I'm physically organized to know, seemed ascendant to me, and a beautiful reminder of what is most profoundly humble about being human. I was told the comma looked like a typographical error, so I removed the comma, and then the title of my first full-length solo work read like an expression of outrage.

In a later solo show, I asked the question: Is consciousness a body part? It's a scientific question, a theological question and an ontological question, but for me, mostly, it's a performer's question, and the answer, discovered through too much talking and finally silence, finally dance, finally stillness in this piece, was

Yes.

The rehearsal lab

We spoke to five contemporary theatre artists about their work, the ideas in their work, and how they think about reaching their actors and their audiences. These are a smattering of their responses.

Mallory Catlett is an Obie-winning director and theatre devisor in New York City (NYC). Her work has toured to American Repertory Theater (Cambridge), the Adelaide Festival (Australia), the Brighton Festival, Bristol's Mayfest (UK) and Brooklyn Academy of Music's (or BAM's) Next Wave Festival (NYC). Her work has also premiered in NYC at The Ohio, HERE Arts Center, The Ontological-Hysteric Theater and PS122, among others.

Making an idea operational:
I come from being a dramaturg, I like theory, I like philosophy, I like the theoretical aspect of things, of content in any way. In Richard Foreman's work, he says, 'It's not about the expression of an idea, it is how to make an idea operational in the creation of the new.' The idea of how to innovate or to make something new does not have to do with 'I have something unique to express' or 'I have unique ideas.' I don't. I'm interested in how I can use the theatre to make ideas operational, in which case, they create the new. ... I try to take very large philosophic ideas, and make them as concrete as possible in the theatrical world.

Reckoning:
I long for a way to live in time differently than I do. I'm very influenced by things like quantum mechanics, where it's not our reality. I mean, we

live in a very causal reality; there are circumstances when this causes that, and this causes that, but that's just simply not true for the way the universe works. I think we live very out of sync with the natural world, and that causes tremendous anxiety. ... [Proust] changed what literature could do and was, and people's perception of time. But I'm a theatre artist, and he's already written a novel, so how do I do that within my medium? What interests me, is through all this process of research, how I might be able to actualize it. ... I think ultimately the theatre is always about the present. That's what it's best to do. For me, theater is a great place to reckon with the past in the present.

Christian Parker is a director, dramaturg and consultant. He is the chair of Columbia University's Theatre Program and previous associate artistic director of the Atlantic Theater Company. He has directed and developed works at Sundance Theatre Labs, the O'Neill Playwrights Conference, the Lark, Merrimack Repertory Theatre, Keen Company, P73, the LAByrinth Theatre Company and Soho Rep, among others.

Knowing the audience is following:

I enjoy it when a moment that I've selected – a line or a look or a gesture – gets clocked by the audience, which then pays off down the line of the story. So if you get a reaction – like a laugh or a gasp or you feel like the audience is leaning forward in the moments you want them to. Then that satisfies me. It doesn't matter to me whether you get a standing ovation or whether the audience is the most raucous. But you can tell whether the audience members are following it the way you want them to. Not just what's happening on a literal level, but also if you create a little bit of comic business with a piece and an hour later it gets a big reaction the second time because they clocked it and you wanted that to happen. That's good. That also means that they have been with the story for that intervening hour. It means that they still care.

Layers of complexity:

A lot of dramaturgs will say: I've got to make sure the playwright has been clear about what's happening moment to moment, that the narrative is clear, that the story is clear, that the point is clear, and there is a place for that kind of conversation. But I actually don't think that's the job of the director, or a good dramaturg. It's actually to come to a place of agreement about what the essential story is and then you bring your skills and you capitalize on the actors' and the designers' skills to reveal layers of complexity.

Catherine Fitzmaurice developed the Fitzmaurice Voicework® to help artists reduce tension and improve vocal expressiveness. She has taught all over the world, and has held teaching and consulting appointments at the Royal Central School of Speech and Drama, the Juilliard School's Drama Division, Yale School of Drama, New York University, Harvard University, the Moscow Art Theatre, the Stratford Shakespearean Festival, the Guthrie Theatre and Lincoln Center, among others. She has also presented her work internationally at major medical and theatre conferences, including the 'Freedom & Focus' international conferences on Fitzmaurice Voicework® in Barcelona, Spain, Vancouver, Canada and Bogota, Colombia. For further information, see her article 'Breathing is Meaning' available at www.fitzmauricevoice.com.

Training and science:
The premise of the work I teach – Destructuring/Restructuring – is based upon neurological principles. My interest in the differences between breathing as a survival reflex and breathing for phonation as an intentional act, starting in 1965, led me to explore the autonomic nervous system (ANS) and the central nervous system (CNS), and to find ways to privilege the former in the awareness of the actors I was teaching, rather than other forced and controlled methods of voice training. Learning to flow with the ANS seems to help with favoured qualities in performance such as creativity, spontaneity, intuition and presence. The particular methods I have adapted and developed specifically for the voice are partially available in other fields, such as psychotherapy, body therapies, movement, meditation and mindfulness, all of which have extensive research to support their efficacy – but are mostly silent and directed towards individual development. My work in Voice directly impacts the performer or professional voice user in the very act of communication via language. The work is also informed by my interest in Asian philosophies and arts.

Tristan Sharps is the artistic director of Dreamthinkspeak, an internationally recognized creator of site-responsive performance. He has created works in a former Co-op department store in Brighton, an underground abattoir in Clerkenwell and a disused paper factory in Moscow, among others. His *One Day, Maybe* (2013) was inspired by the Gwangju Uprising in South Korea and was created and first performed in a large disused school. His *The Rest is Silence* (2012) is a deconstruction of *Hamlet* commissioned by the Brighton Festival with London International Festival of Theatre (LIFT) and the Royal Shakespeare Company (for the World Shakespeare Festival).

Space and ecology:
Dreamthinkspeak does a lot of work in old abandoned buildings, where we take over the space and audiences can roam, uncovering the performance. The Hamlet piece [*The Rest is Silence*] was slightly different. Once they [the spectators] were in the space I wanted to give them a feeling – to inhabit that world of Elsinore. On the one hand, there's a bit of claustrophobia and you can move around a little bit. But it's not a vast space. Also there's that sense of voyeurism: you are looking in at people in their private space. They can't see you (and actually the actors couldn't see the audience during their scenes), but you can see them. I don't think it's a very comfortable feeling for the audience. You are spying, you are peeking in on their space. You are also aware of the audience – of being a little bit a part of the show. The other part was the mirror reflections; your image gets picked up and there's a sense of things being shiny on the surface. Wrapped up together, a feeling is created that, they might not be able to articulate, rather than a cerebral awareness: a feeling of being in that space. The sense or emotion of the play.

The performances where you [as audience] wind your way around a building, opening doors, puts you in a very disorientating place. I do this workshop early in rehearsals where everyone picks a place around the room and I tell them to choose a point on the opposite side and walk to it. Then another point. Once they've done that and they return to their original point, they have a pattern. Then they put on blindfolds and try to repeat the pattern. They suddenly are moving very differently, slowly – there is complete disorientation. As soon as you lose orientation, you are in a very interesting headspace. Sometimes there is adrenaline or nerves. You don't know what's happening. That's the feeling the audience gets. They are disorientated and open, so small things can be huge events. Because you have a very sensitive canvas, you don't need a whole splurge of paint. Small things can be huge.

Intuition:
How I approached *Silence* was a very intuitive choice. There was a sense in the play of time being suspended so I wanted to shine a light on that. ... I pursued it very rigorously in terms of how I was going to work with it in the text but I didn't pursue it in terms of a clear cerebral conscious map. It was very intuitive in terms of, 'I knew that had to happen', I knew that everything had to somehow relate to everything

else and that the audience would get some of it and not all of it and that's fine and a different audience would get different bits of it. By somehow dissolving ... I wanted everything to dissolve a little bit, like the whole structure of the play would almost dissolve. ... I don't think I have the analytical machinery to make a clear map of why I did that at every single moment in the play, I just had an intuition that that's what I had to do. And I had to be bold with it as well. And also trust that out of that I would find the right structure.

Language: The rhythm of the language is key.

Tim Miller is a performance artist who explores the artistic, spiritual and political topography of gay identity. His pieces include *My Queer Body, Glory Box* and *1001 Beds*. He has performed in such venues as Yale Repertory Theatre, the Institute of Contemporary Art (London), the Walker Art Center (Minneapolis) and BAM. His book *1001 BEDS* won the 2007 Lambda Literary Award for best book in Drama-Theatre. He has taught performance at University of California, Los Angeles (UCLA), New York University (NYU), the School of Theology at Claremont and at universities all over the United States. He co-founded Performance Space 122 in New York City and Highways Performance Space in Santa Monica, California. He has received numerous grants from the National Endowment for the Arts. In response to our request, Tim wrote the following short performance piece:

'Heart on My Sleeve: A Theater is the Opposite of a Cemetery'

(Stage Directions: The performer – with a stethoscope around his neck – slowly goes to each and every reader who will ever read this essay and listens to their heart. Intoning while he does so some of the metaphorical, or maybe MEATaphorical is better, images of the heart: Cold-Hearted, Warm-Hearted, Lion-Hearted, Bleeding-Heart, Broken-Hearted, Heart on My Sleeve.)

I am a performance artist who traffics in the poetics of the body.

In my own performance work – and the extensive teaching that I do at universities and arts centres – I sometimes feel like a bit of a doctor as I help prompt an auto-ethnography of flesh and bone and ask folks to dig deep into that overlap of memory, desire and history that constellates in every body. I have invited many thousands of people around the world to spelunk into an exploration of the narratives and metaphors that live in their embodied self – the collection of cells, feelings, race, gender, soul, wounds, family narrative – that assemble into a person. I freely admit that I have more than a little bit of that familiar 'Creative Type' trust – even valorization – of metaphor and imagination that may sometimes seem contradictory to a Western cognitive and scientific model. But that binary has never really held up for me though; those graceful spiralling tendrils of DNA are a crucial part of our mythopoetic family tree at the heart of how we understand ourselves. Just as I am interested in where the poetics of the body smacks up against societal concrete systems, I am intrigued with where metaphor collides with science because this is where the rubber hits the road.

Just as I might ask participants in one of my workshops to locate a charged narrative of a single place on their bodies, let me tell you a story.

I saw my heart on my birthday not long ago. A magic wand, okay an echocardiogram ultrasound wand, showed me my heart on my birthday. Echocardiography uses Doppler ultrasound to create images of the heart and its size, shape and pumping capacity. I had just performed on a college campus in a theatre that was named for the cemetery where most of my family on both sides are buried or were cremated. I had never performed in a theatre named for a cemetery – and especially not one so linked to my family – so this quickly brought up some anxiety. See I don't want to think a stage is cemetery, a theatre a crypt, it is supposed to be a marriage bed or a place of birth. A theatre is NOT a cemetery! But nonetheless, my heart started hurting.

(Stage Directions: Enter DNA stage left.)

My Dad's heart had hurt. He died of a heart attack when he was just a few years older than I am now. I became anxious that my heart was breaking like my dad's heart had broken and they cremated him at the cemetery that gave the money for the theatre I had just performed in.

My Grandfather's heart had hurt. He had fled the family farm in Kansas and come to LA where he became a housepainter until he died of a heart attack in his mid-60s and he would be buried in that same cemetery-cum-theatre.

My Great-Grandfather's heart had hurt. He had driven the Model T from Kansas to LA in 1921 and he was so upset on 7 December 1941, after the attack on Pearl Harbor that he had a heart attack and drove his car into a light pole on the streets of LA and died there in traffic in his mid-60s.

Do we see a trend? You can imagine why performing in a theatre named for my family cemetery might trigger a perfect storm of anxiety! So I went to the doctor and I wanted him to go inside my heart – inside my queer heart – and see what was there … to see if it was breaking. I did not go to a performance art psychic surgeon; I went to the cardiologist.

(Stage Directions: Enter avuncular medical technician.)

At the Venice Beach cardiologist I took off my shirt and the Russian technician Yakov lubed my heart up and the cold wand touched my skin for the echocardiogram. He showed me the monitor and I saw my heart throbbing and sucking there in the ultrasound murk. It looked like a night-vision Xtube gay porn scene.

Yakov asked me, 'What does it look like?' as he pointed at my aorta.

Oh no, a test! To me my aorta looked like a puckering anus, but I knew that wasn't the right answer.

'Um.' I bought some time.

Yakov tried to give me a hint and made a fish mouth with his lips and I knew what we has in fact fishing for.

'A fish', I said in Double Jeopardy triumph and he beamed at me. What a good student am I!

'Yes, aorta looks like fish mouth!' Yakov exclaimed. You can never escape metaphor!

I stared there at my fish-mouth heart. My anus-heart. My fish-mouth heart ... thirsty, gaping, gaping, hungry, breathing, beating all these fifty-six years. It was like looking down the branches of my family tree that connect me to Kansas in 1871, Pearl Harbor Day, and to sitting with my dead Dad's body before they cremated him at that Cemetery with its name on a theatre.

Everything seemed to be okay for now – to be confirmed later on by the cardiologist – and my new Russian friend sent me on my way. I put my shirt on and then paid my ridiculously high co-pay for my next-to-worthless health insurance that would cover none of this procedure. I know, I am going off subject with this political aside, but here's a theorem ... Poetics + Science = Politics!

(Stage Directions: Time for the 11 o'clock number! Followspot on Tim.)

A theatre should not be named for a cemetery. A theatre is the opposite of a cemetery. On my birthday science and art collided and I saw my heart thirsting, fish-mouth, anus-heart thirsting. My queer heart thirsting. What is a queer heart? Maybe it's light in its loafers. Perhaps has a limp-wristed aorta. Maybe it eventually does skip a beat down the garden path.

I am not my Father's heart.
I am not my Grandfather's heart.
I am not my Great-Grandfather's heart.

I know my heart will stop one day– we all day of a stopped heart –but not today on my birthday.

Today my heart will love.

Today my heart will send that thirst and rhythm out into the Pacific Ocean when I have my swim.

Today my heart will dare to imagine that things can get better in this heavy-hearted world.

Today my heart will spin poetics and science into a bouquet of delicious arterial cotton candy.

Today my heart will look getting older and dying direct in the ultrasound magic eye.

I unlocked my bike and rode down to the ocean in Venice Beach so I could buy my husband Alistair some sushi for our lunch and pedalled slowly home. My heart beating with each radiant turn of the bicycle wheels.

Appendix

Abstracts of a Few Influential References

For further reading

Sources from the Languages section:

Mandler, Jean M. and Cristobal Pagan Cánovas, 'On Defining Image Schemas'. *Language and Cognition*, Vol. 6 (2014), pp. 510–32.

This paper revises the concept of image schemas – one of the staple tools in cognitive science and cognitive linguistics. Originally, the schemas were defined by Mark Johnson and George Lakoff, and presented in their numerous publications in the 1980s and 1990s. Image schemas were then defined as basic concepts of spatial organization and force (in the sense of acting upon objects, or being acted upon) and were claimed to underlie more complex conceptual patterns, such as conceptual metaphors. The difference was that while conceptual metaphors were defined as mappings between two rich structures, image schemas were not mappings and provided only a skeletal background to those more complex relations. Among others, it was claimed that the two domains involved in a conceptual metaphor need to share some image schematic structure to be used in a mapping. Mandler (Department of Cognitive, Science University of California San Diego) and Cánovas (Institute for Culture and Society, University of Navarra) build their approach by looking at actual stages of early child development, when some schemas are observed early on, while others take more time – which supports other criticisms raised and provides evidence for their claims.

They argue that image schematic concepts develop gradually, starting (in the early months of an infant's life) with the simplest ones, called *spatial primitives* (such as PATH, LINK, THING, CONTAINER, APPEAR, etc.), which can be combined into *image schemas* (as in MOVE THING or OPEN CONTAINER). The paper is very important for our understanding of complex schematic integrations and also of blends emerging later in life. It offers a clear trajectory for the emergence of complex concepts and preserves their embodied grounding while clearly distinguishing the ways in which embodiment triggers conceptualization. It gives texture to many simplistic claims about the role of embodiment in conceptualization of events and emotions.

–Barbara Dancygier

Hutchins, Edwin, 'Material Anchors for Conceptual Blends'. *Journal of Pragmatics*, Vol. 37 (2005), pp. 1555–77.

This paper makes very important connections between postulated conceptual structures, such as blends, and material phenomena, often of specifically cultural nature. The postulated role of material objects is independent of general mechanisms of representation; instead, Hutchins (Department of Cognitive Science, University of California, San Diego) proposes that material structures are inputs to other blends, independent of other participating concepts. Blends that rely on material inputs, and thus have material forms as their elements, have, in Hutchins's terms, increased stability. Hutchins argues further that material aspects of the world do not need to enter blends through an independent step of their conceptual representation – they participate as material inputs.

Hutchins discusses several examples. His primary case is the concept of a queue, which takes a material form of bodies arranged in a line, but relies centrally on the idea of sequential order, rather than just a line (not every line of bodies is a queue). As a blend, the queue constructs a sequence of persons, but also gives the blend the required stability, so the structure of the queue is fixed while the sequence is realized through the motion of people in the line.

As Hutchins also points out, cognitive work can be done through imaginary manipulation of a physical structure. This provides an elegant explanation of how essentially material structures may persist through time. Even when the objects themselves are not present, the conceptualization persists or even becomes independent of them. Hutchins argues that much of cognition (in areas such as logic or mathematics) is not purely conceptual, but has built on stable material blends.

As Hutchins also points out, these phenomena are systematic, not just isolated 'clever cognitive strategies'. He also argues for the concept of 'an ecology of conceptual blends', as an accumulation of complex properties. For example, one does not need to be aware of the whole history of a tool to understand how it works today; the stable conceptualization initiated through a blend with a material form is sufficient.

Hutchins's work is very important to all forms of art which rely to some degree on materiality. Theatre is definitely such a form – to give just one example, the concept of a stage as a blend of material location and narrative meaning is among the most stable ones, while the very material forms of theatre may have changed. The concept persisted through time thanks to its material grounding, but became an element of an ecology of blends that theatre practitioners use daily.

–Barbara Dancygier

Bolens, Guillemette, *The Style of Gestures: Embodiment and Cognition in Literary Narrative*, trans. Guillemette Bolens, Baltimore: Johns Hopkins, 2012, first published in French, 2008.

The Style of Gestures is an exercise in comparative literature, ranging from Proust to James Joyce, as well as a work which applies neuro-scientific findings about the brain to the acts of reading and interpreting literary texts. Bolens's central thesis is that as readers we understand similes, images and metaphors in literary texts through 'kinesic intelligence'. Drawing on the work of Ellen Spolsky, Bolens defines kinesic intelligence as the ability to use our own experiences and memories of bodily movement to interpret other people's gestures in real life and as they are depicted in visual art and literature. Kinesic intelligence is thus both proprioceptive, helping us to be more aware of our own embodied processes of understanding, and integral to forging connections between our body and the bodies around us. In providing a framework for how we use our experience of moving our bodies to anticipate how real and fictional situations will pan out, kinesic intelligence has strong links with simulation theory in cognitive science, which also seeks to explain how the embodied brain can hypothesize about events.

–Laura Seymour

Gibbs, Raymond W., 'Metaphor Interpretation As Embodied Simulation'. *Mind and Language*, Vol. 21, No. 3 (2006), pp. 434–58.

Conceptual metaphors are often described in terms that suggest they are abstract, purely cognitive entities: schematic mappings from one domain of knowledge to another. Psycholinguistic research supports the claim that the underlying basis for metaphorical language and thought is often grounded in physical experience – for example, metaphorical expressions about abstract desire across languages are constrained by speakers' specific intuitions about their physical experience of hunger. There is also broad support for the claim that language about literal, concrete events prompts people to simulate some experience of those events. Here, Gibbs argues that understanding metaphorical language in the moment also relies on imaginatively simulating embodied experiences of the described activities, even if those activities are physically impossible. In a series of experiments, Gibbs and colleagues found that while people reported that it was easier to form images for non-metaphorical sentences, their mental images for metaphorical expressions like *grasp the concept* or *coughing up a secret* were in fact detailed, dynamic, multisensory, kinaesthetic, and enhanced by either performing or witnessing the relevant embodied actions (e.g. grasping, coughing) beforehand.

–Vera Tobin

Talmy, Leonard, 'Fictive Motion in Language and Ception', in P. Bloom (ed.), *Language and Space*, Cambridge, MA, MIT University Press, 1996, pp. 211–76.

The phrase 'fictive motion' refers to expressions that use verbs of motion to describe scenes involving no actual physical movement, such as *The road goes over that hill*. This essay coins the term and presents a detailed description of the phenomenon. The technical use of 'fictive' suggested here is opposed to 'factive' and means something different from 'fictional'. The idea is that, regardless of whether a given sentence refers to a fictional or non-fictional event, that event can be described in terms that are more or less veridical. Thus, a factive motion use of 'run' would describe literal running, as opposed to the fictive motion of a sentence like *The cliff wall runs along the coast*. This notion of fictivity has been influential, and is now commonly used in linguistics. Fictive motion in particular has attracted attention in cognitive science, following the question of whether or not these descriptions of physically static scenes nonetheless evoke dynamic mental images and simulated scanning over the field of vision. See especially the work of Teenie Matlock and collaborators (e.g. Matlock 2004; Richardson and Matlock 2007) for evidence that they do.

–Vera Tobin

Barsalou, Lawrence W., 'Perceptual Symbol Systems'. *Behavioral and Brain Sciences*, Vol. 22, No. 4 (1999), pp. 577–660.

This influential article argues against the classical view that knowledge is represented in cognition in amodal (that is, abstracted away from any particular sensory modality) data structures that are processed independently from the brain's modal systems for things like action and perception. Instead, Barsalou proposes, knowledge is grounded in a system of 'perceptual symbols'. These symbols are not distinct from perceptual processing. Instead, they are records of neural states that underlie perception. Specifically, they are the records of the subsets of those neural states that are shared across many separate instances of related perceptual experiences. These shared pathways are reinforced by their co-activation. That partial nature is what makes perceptual symbols schematic, and the reinforcement is what constitutes the record – it's not stored separately in an abstracted form, but rather is directly instantiated in these neural traces. Much of Barsalou's work following the publication of this piece has pursued the details of this theory, finding, for example, that food concepts seem to be represented in part as simulations in the gustatory system (Simmons, Martin and Barsalou 2005).

–Vera Tobin

Sources from the Bodies section:

Libby, Lisa K., Thomas Gilovich and Richard P. Fibach, 'Here's Looking at Me: The Effect of Memory Perspective on Assessments of Personal Change'. *Journal of Personality and Social Psychology*, Vol. 88, No. 1 (2005), pp. 51–3.

Lisa Libby's work principally centres on the affects that occur when individuals take either a first- or third-person perspective of themselves. People who take a first-person perspective when remembering an event from their life are more likely to recall specific, concrete details. However, taking a third-person perspective encourages individuals to take more abstract, 'big picture' meanings. This affects not only how we remember the past but also how we speak about ourselves and events and other behavioural traits. For example, people might express that they 'feel like a different person' after a major life event. Libby and her fellow researchers suggest that undergraduates often enter college as a means to reset their identity and leave behind a former image. In one study, the researchers recruited students who self-identified as socially awkward in high school. The subjects were asked to remember and rate an awkward event in high school and imagine it from either a first- or third-person perspective. They found that when the subjects imagined themselves from the third person they felt like there were broader changes in social behaviour and, when confronted with a confederate whose job it was to rate their present social interaction, they were more confident and outgoing. Their sense of a newer, better self was reinforced by the shift in perspective. For an individual to perceive of such a dramatic change she has to be able to visualize herself as one thing and then another, presumably side by side, in order to conceive of herself as different.

–Neal Utterback

Neal, David T. and Tanya L. Chartrand, 'Embodied Emotion Perception: Amplifying and Dampening Facial Feedback Modulates Emotion Perception Accuracy'. *Social Psychological and Personality Science*, Vol. 2, No. 6 (2011), pp. 673–8.

Neal and Chartrand's paper proposes that automatic facial mimesis positively contributes to emotion perception, as demonstrated through twinned emotion perception experiments. Subjects in both experiments took Baron-Cohen and colleagues' 'Reading the Mind in the Eyes Test'. Experiment 1 suggests that emotion perception may be hindered when facial feedback signals are dampened, following the testing of patients who had recently received cosmetic Botox injections. Meanwhile, Experiment 2 suggests that emotion perception may be enhanced when facial feedback signals are enhanced following the application of a constricting gel, tightening subjects' skin to strengthen afferent feedback from muscular movement. These experiments support the understanding that we automatically mirror others' facial expressions, and further suggest that we implicitly rely upon sensory feedback from our own facial muscles to accurately interpret others' emotional states. Altogether, the paper demonstrates that empathic mirroring plays a significant role in moderating emotion perception, which is important for successful social cognition. For the purposes of performance theorists, this paper insists that we attend to the embodiment of spectators and acting partners alike when considering how each interpret the emotions or intentions of other bodies onstage.

–Christopher Jackman

Sources from the Ecologies section:

Gallagher, Shaun, 'The Socially Extended Mind', *Cognitive Systems Research*, No. 25–26 (2013), pp. 4–12.

Gallagher defends a liberal interpretation of the extended mind hypothesis, which views cognition as 'support[ed] and extend[ed]' (4) by the practices of social and cultural institutions. Gallagher defines cognition as 'an enactive and emotionally embedded engagement with the world', (11); this cognitive engagement is aided by a variety of social practices and institutional procedures, or 'mental institutions', that extend the processes of cognition beyond what the narrowly defined mind might be able to accomplish alone.

These mental institutions include 'cognitive practices that are produced in specific times and places' and that are 'activated in ways that extend our cognitive processes when we engage with them' (6). The legal system provides an example of such an institution, in which case law can be accessed external to the individual brain in ways that accomplish some of the cognitive work of reaching judgements. Our engagement with mental institutions, as with technologies or tools, extends our cognitive processes beyond their innate capacity.

Gallagher reviews foundational literature on the concept of the extended mind (Clark and Chalmers 1998), as well as significant objections to the concept. He asserts that an enactive approach rather than a functionalist approach to extended mind theory provides a more meaningful defence against the major objections that have been raised to the idea.

If the extended mind is a socially extended mind, Gallagher suggests in his conclusion, then significant social and institutional practices must be interrogated for their assumptions and biases. The extended mind theory thus offers a tool whereby critical theory might examine ways in which 'our communicative practices, our possibilities for action, our recognition of others, our shared and circumscribed freedoms' (12) are impacted by these embedded praxes.

–Sarah McCarroll

Landau, Ayelet N., Lisa Aziz-Zadeh and Richard B. Ivry, 'The Influence of
   Language on Perception: Listening to Sentences About Faces Affects the
   Perception of Faces'. *Journal Neuroscience*, Vol. 30 (2010), pp. 15254–61.

Landau et al.'s paper provides preliminary evidence of the ways in which linguistic descriptions may prime or affect visual perception. The paper explores the complex relationship between linguistic and non-linguistic representations of the world, aiming to further the investigation of the ways in which the younger system may 'piggyback' on the more ancient. Landau et al. outline a history of studies into this interaction between fundamental and high-level processing of the environment, but they note that the mechanism(s) through which they might interact are far from established.

In their study, the research team tested the effects of visual and linguistic priming on an established early response to faces in EEG data. In one experiment, linguistic or visual primes of faces or scenes were followed by probe images, again of either faces or scenes. Participants then responded by indicating the plausibility/implausibility of the primes and probes (either spoken sentences or visual images could have unusual features such as very long eyelashes or upside-down mountains).

Following a series of three experiments based around similar methodology, Landau et al. found that language *disrupted* systems related to perceiving faces, slowing down response rates and minimizing the magnitude of EEG responses. In their discussion they suggest that there may be competition between linguistic and visual processing and the processing of immediate (visual) and displaced (described) content – language may affect perception without necessarily improving it.

–Matt Hayler

Lupyan, Gary and Emily Ward, 'Language Can Boost Otherwise Unseen Objects Into Visual Awareness'. *Proceedings of the National Academy of Science of the United States of America*, Vol. 110 (2013), pp. 14196–201.

Lupyan and Ward's paper is situated in an extensive (and contested) literature regarding visual perception's immunity/susceptibility to external influences like language.

In their research, Lupyan and Ward supressed participants' vision of a familiar object through the established technique of continuous flash suppression. They found that the introduction of a correct verbal label, for example, saying 'chair' when the participant's vision of an image of a chair was being supressed, was able to boost the object into perception; that is, correct verbal labels heighten the perception of otherwise supressed objects. Results were significant against no verbal label, and incorrect verbal labels decreased performance.

In their discussion, the team propose that though 'it may seem maladaptive for vision to be so sensitive to input from outside its "domain" …, if we consider that the real purpose of perceptual systems is to help guide behavior according to incomplete and underdetermined inputs, and that perception is at its core an inferential process, then perception needs all of the help it can get'. If the aim of perception is to produce an objective representation of the world then vision's susceptibility to external influence would seem, as Lupyan and Ward suggest, counter-intuitive. But if the aim is to act successfully, then a more complex interconnection of systems should be expected.

–Matt Hayler

Notes

Introduction

1 Some of this section is derived from Rhonda Blair and John Lutterbie, 'Introduction: *Journal of Dramatic Theory and Criticism*'s Special Section on Cognitive Studies, Theatre and Performance', *Journal of Dramatic Theory and Criticism,* Vol. XXV, No. 2 (Spring 2011), pp. 61–70.
2 In addition, a growing bibliography and resources can be found at http://performancesciencecreativity.com.

Chapter 1

1 Examples of Evol's work can be found at http://www.nimball.com/evol.html.
2 An image can be found in the following website: http://www.artfido.com/blog/the-most-creative-sculptures-and-statues-from-around-the-world/, image numer 4.
3 A good overview of the issues can be found in Fillmore (1997); for a discussion of cultural specificity of deixis, see Hanks (1990). Deixis in narratives is broadly discussed in Duchan, Bruder and Hewitt (1995).
4 This concept is further complicated by Weimann's (2000) division into locus and platea.
5 I use the term 'blend' on the basis of the theory of conceptual integration, proposed in Fauconnier and Turner (2002).
6 For a discussion of these divisions from an historical perspective, rather than a linguistic one, see Robert Weimann (1987).
7 All references to Shakespeare's plays are from: Shakespeare, William. 1994. *The Complete Works.* The Shakespeare Head Press Oxford Edition. New York: Barnes & Noble Books.
8 In an earlier work, Dancygier (2012), I discussed one of these forms in some detail, as a linguistic construction used in poetry and on stage.
9 Much of the current understanding of the role of the body builds on the theory of embodied cognition. A number of scholars Johnson (1987 and 2007), Gallagher (2005), Prinz (2002 and 2004), Gibbs (2005) and Shapiro (2010) now follow the general assumption whereby much of human cognition, including the use of language, understanding of emotions, etc. is centrally rooted on embodiment.
10 Elsewhere cf. Dancygier (2010), I discussed the speech from the perspective of its use of negation; the reader may also be interested in the work on the use of gesture in *Julius Caesar* (Sweetser 2004, Seymour *this volume*).

Chapter 2

1 Thanks to Amy Cook, Rhonda Blair, Gillian Woods, Laura Salisbury and Isabel Davis, and to Raphael Lyne and the other participants of the Renaissance kinesis workshop in honour of Guillemette Bolens at Cambridge on 25–27 September 2014, for their suggestions on this chapter.

2 First published in 1623 in the First Folio, the play was probably first performed in 1599. The tourist Thomas Platter's diary records visiting the newly opened Globe around 2.00 pm on 21 September 1599 to see *Julius Caesar*.

3 North's conspirators simply 'press' close to Caesar.

4 Oxford English Dictionary, online version, 'petty', *adj.* and *n.*

5 For instance, Gervase Babington (1592: X7ʳ), who would become Bishop of Worcester, states of 'kneeling & bowing', 'outward gesture dooth helpe our inward heart, and stir us vp rather to reuerence'. Ramie Targoff provides a good discussion of these ideas (2001: 4). The earliest discussion of kneeling in the West, by Peter Cantor in the twelfth century, focuses on the cognitive effects of kneeling, see Trexler (1987).

6 Oxford English Dictionary, online version, 'boot', *adj.* 1; 'boot', *n.* 1.

7 She continues, 'In so far as the influence of habit causes a translation of soul into body and body into soul, these two will form a unity-in-separation, an absolute unity without fusion. ... Between container and contained, a reversible relation abolishes the partition between exterior and interior, allowing soul – henceforth constituted as "Self" – to relate to the world, the real externality', relating to the world in a way that reflects the world back into self-consciousness.

8 See for instance, Lutterbie (2011).

9 In the prompt book for this production, scene 3.1 is labelled 'THEN FALL, CAESAR', in pencil, suggesting that this image of falling was central to the cast's interpretation of the play Rylance (1999b: 31).

10 For instance, the 5th move is as follows: 'Caesar grabs Metellus right arm. Caesar circles with Metellus, holding Metellus' sword arm, and points sword' Rylance (1999a).

11 Michael Billington, 'Saturday Review', *Guardian*, 29 May 1999: 4.

12 This theatre, opened in September 2014, is a reconstruction of what is believed to be the first Shakespearean theatre to be built in continental Europe during Shakespeare's lifetime. The original site housed travelling players from England who performed classics (including Shakespeare's *Hamlet*) to Baltic theatre lovers.

13 Lloyd (2013: 18) analogizes the lack of freedom in contemporary prisons and the way in which 'for the conspirators, Caesar represented an erosion of fundamental civil rights so huge, so towering, terrifying and confining that the conspirators believed they were in a prison'. Harriet Walter

(Brutus) draws the same comparison in an 'Interview' Lloyd (2012: 27): 'The metaphor of an incarcerated group of people who are dependent on favours and handouts and punishments and everything else from a superior power is also neat. That is everyday life in a prison.' Lloyd's 'Production Diary' demonstrates how seriously the prison setting was taken: during rehearsals, 'staying true to our locked prison setting, no-one leaves the space. This means that in the first run, props go astray, wires end up in a tangle and costumes are left all over the place,' Week Six, 20. The actors used only those props and costumes that would have been available to prisoners. For instance, the phrase 'beware the Ides of March' is found in a magazine horoscope.

14 The audience also experienced the death physically; the conspirators jostled the audience as they crowded to force Caesar to drink bleach (perhaps revenge for Caesar humiliating Cassius by force-feeding him a doughnut because, in a scene that drew upon contemporary discourses of shaming thin women, he was 'too thin'), before stabbing him. *Julius Caesar* (2013).

Chapter 3

1 Available at the McGraw-Hill 'Online Learning Center': http://highered. mheducation.com/sites/0072831820/student_view0/glossary.html.

2 Indeed, cognitive linguistics has a rich history of exploring linguistic structures that encode and reflect elements of personal viewpoint, from Charles Fillmore's work on deixis (1975) to Len Talmy's work on attentional windowing (1996 and 2000), Ronald Langacker's work on dimensions of construal (1993) and Brian MacWhinney's work on the emergence of grammar from perspective (2005). More generally, the field's emphasis on the embodied nature of human thought, see Lakoff and Johnson (1999), Gibbs (2006), Dancygier and Sweetser (2005), has helped to ensure that viewpoint gets its due.

3 For more detailed discussions of the compressions involved in this blend of living actor and performed character, see Cook (2007 and 2010) and McConachie (2008).

Chapter 4

1 I discovered an interesting phenomenon as we went through our third-person mental imagery. One of my actors would use her index finger like a kind of third-person finger puppet to imagine moving through the space. When her finger puppet arrived at the prop table, the actor would adopt first-person gestures and mime putting on or taking off a wig or physically

moving a chair. She would then return to her third-person finger-puppet self. The actor had previously self-reported that she often found strict blocking and choreography difficult to execute. However, I observed an increased level of confidence once we added the mental imagery sessions. She seemed to be able to add another level of comprehension by imagining herself moving through the space and performing the actions on cue. My previous empirical studies have shown that actors who have free use of their hands can encode and retrieve text more successfully than those whose hands are restricted or choreographed arbitrarily. During our text speed drills, the actress in my production would switch from a first-person gestural perspective – for example, she would mime putting on a wig – to a third-person perspective – for example, using her index finger to 'move' across the stage like a finger puppet. See also Cook and Goldin-Meadow (2006), Frick-Horbury and Guttentag (1998) and Utterback (2013: 50–87).

Chapter 6

1 Because I address theories of flow below, I should note that flow theory's own major theorists have approached the topic of acting with fraught misconceptions about its overlap with social 'skill sets'. Sawyer, for instance, suggests that the actor's skill lies in her ability to make both scripted and unscripted dialogue sound 'natural' by attuning the micro-timing of conversational rhythms (2006: 245, 247). While wholly ill-suited to explicitly unrealistic expressive forms, Sawyer's misconception remains woefully inadequate even for 'realistic' styles of performance, confusing a more or less relevant effect for a criterion of excellence and tacitly dismissing actor training as inferior to a practice of conversation. I trust that my discussion below provides a more nuanced understanding of both skill and flow in performance.
2 Here, I refer to specialized training programmes that demand a precise, highly technical control of one's body and awareness, and that are based upon a clear programme of technical interests, including Meyerhold's system of biomechanics or the Alexander technique (though such examples are unquestionably subject to an evolution of re-emphasis and reinterpretation).
3 While this model of consciousness has been previously conceived as a dual *system* theory, Stanovich's change in vocabulary, from *system* towards *process*, reflects an emphatic recognition that these distinct *processes* do not emerge from distinct neurological and cognitive systems. The dual process theory allows us to forge functional distinctions while recognizing a shared architecture of mind. See Evans (2012).
4 According to a series of experiments cited by Jeannerod, our awareness of an impulse to move typically precedes the action itself by about 206

milliseconds, but the neural preparation for that action precedes our awareness by an additional 345 milliseconds (2006: 60–1).

5 Stanovich suggests that when meaningful errors do occur in the typical function of well-practised activities, a process called override detection is triggered, where action emulation is employed to identify and draw Type 2 attention towards the error (Stanovich, West and Toplak 2011: 366). For example, we may not be aware of our minute responses to the terrain as we navigate the sidewalk, or of a habitual tendency to drag our feet, but this information rushes to the fore of our attention if we stumble on a flagstone.

6 Grezes and colleagues emphasize the role of canonical neurons in these processes, which are neural structures that respond consistently to object-specific actions and perceptions (2003: 928). Kohler and colleagues' primate study (2002) is noteworthy for recognizing that the relative vibrancy of percepts in action observation is heavily contingent on the salience of that percept to the completion of the act itself. For example, a monkey that witnesses a nut being broken will physically and neurologically anticipate its own performance of a breaking action, but that response will be stronger in response to the sound of breaking, as this is a more tightly coupled index of the action, see also Aglioti and Pazzaglia (2010), Ricciardi et al. (2009).

7 Hogarth tacitly but resoundingly rebuts this assumption in his text, *Educating Intuition* (2001).

8 Limb and Braun's fMRI study of solo jazz pianists (2008) seems to support this link between skilful creation and mindful cognition from the other direction, demonstrating that during improvisation, jazz pianists demonstrated a synchronization of neural processes along with signs of dissociative, transcendent thought.

9 I have had the pleasure of training with Zarrilli in the summer of 2008 and in fall 2011. My experiences are consonant with those described in his writings, and with video documentation of Zarrilli's *kalarippayattu* work archived at the Centre for Performance Research in Aberystwyth, Wales.

10 Supported by an Australian Research Council Linkage Project grant (LP130100670). For further information, contact Kate Stevens at: kj.stevens@uws.edu.au, http://marcs.uws.edu.au, http://katestevens.weebly.com.

Chapter 7

1 For cognitive niche construction, see Clark (2006), as well as Tribble and Keene (2011).

2 As Rick Kemp has shown, such dualist thinking can impede an understanding of the 'psychophysical' underpinnings of the art of acting. See Kemp (2012).

3 For the Yogi Berra, quote, see http://quoteinvestigator.com/2011/07/13/ think-and-hit/. See also Sutton (2007).
4 For a discussion of coaching cues, see Sutton (2007).

Chapter 8

1 Gallagher further expands on this idea: 'In most everyday circumstances, volitional movement means reaching to grasp something, or pointing to something, and the focus is on the *something*, not on the motor act of reaching or grasping or pointing. These movements tend to follow along automatically from the intention' (2005: 49).
2 Our proprioceptive sense provides information about how the body and its limbs are positioned; it is related to kinaesthesia, which is an awareness of the body's relationship to the space surrounding it and objects inhabiting that space.
3 Victoria's reign lasted from 1837 to 1901; Edward VII reigned from 1901 to 1910.
4 This is the period in which the very idea of 'white-collar', that is, intellectually based, jobs emerged, as distinct from 'blue-collar' work, which was defined by physical labour.
5 These ideas were first articulated by Thorstein Veblen in *The Theory of the Leisure Class*. He names them conspicuous consumption and conspicuous leisure. Veblen also includes a third category, conspicuous waste, which applies here, in that clothing was not used until it wore out in high society, but exchanged for new as seasonal fashions changed. Also useful is Quentin Bell's *On Human Finery*, which accessibly articulates Veblen's theories and applies them specifically to dress.
6 For a detailed discussion of the phenomenology of performers and performances on the Realistic stage, see States (1985).
7 In addition to Booth's *Theatre in the Victorian Age* (1991), examinations of Victorian and Edwardian theatre that pay particular attention to the cultural resonances of performance and the situation of theatre within British society include: George Taylor's *Players and Performances in the Victorian Theatre* (1989) , which examines the intentions of actors in performance; J. C. Trewin's *The Edwardian Theatre* (1978), which provides an important reminder that theatrical dynamics were different under Albert than they had been under his mother; *The Making of Victorian Drama* by Anthony Jenkins (1991), an examination of the relationship of Victorian drama to the shifting character of royal life, to increased investment in historical research and production values, and to the changing place of actors and actresses in society; and Powell's *The Cambridge Companion to Victorian and Edwardian Theatre* (2004).

8 This dynamic is not limited to the late nineteenth century; it is the effect of long late-Victorian corsetry, no matter when it is worn. As any contemporary actress who has worn a corset can attest, it is necessary to put on one's shoes *before* being laced into one's corset. If this order is not observed, the actress finds herself obliged to ask for help in getting into her footwear, since her corset will not allow her to bend forward enough to fasten her boots or shoes. Any attempt to lean forward will press the bones of her corset uncomfortably back into the stomach and pelvis.

9 This brief bit of play in the sixteenth century obviously ignores a number of major issues that would have to be addressed in a thorough study of the body map in the period, not least the convention of theatrical cross-dressing.

10 Think of the ways we as a society try to address bodies that are different from the familiar. As a culture, we often struggle to simply accept and view without discomfort bodies that do not conform to our understanding of how the body 'ought' to appear. Amputees, the disabled, those who conform to a 'non-normative' understanding of gender or sexuality; we are in many cases culturally conditioned to look away.

Chapter 9

1 For more on these relationships, see Ihde (1990: 72–112, 2009: 42–4) and Verbeek (2005: 125–8).

2 Parts of this discussion draw on the third chapter of my *Challenging the Phenomena of Technology* (Hayler 2015). The later idea about the grammar of hyperlinks is also developed in an unpublished section of my doctoral thesis, 'Incorporating Technology: A Phenomenological Approach to the Study of Artefacts and the Popular Resistance to E-reading' (2011).

3 Becoming, as described by Martin Heidegger, 'ready-to-hand'. There are two ways in which we can experience objects in the world for Heidegger: (i) in a theoretical stance where we encounter Things which are 'present-at-hand', available for observation, but unavailable to experience as-they-are, and (ii) in use where we can more closely encounter equipment as what it is, as 'ready-to-hand':

> Only because equipment has *this* 'Being-in-itself' and does not merely occur, is it manipulable in the broadest sense and at our disposal. No matter how sharply we just *look* ... at the 'outward appearance' ... of Things in whatever form this takes, we cannot discover anything ready-to-hand. ... The less we just stare at the hammer-Thing, and the more we seize hold of it and use it, the more primordial does our relationship to it become, and the more unveiledly is it encountered as that which it is. (1962: 98)

It is only because things have a reality, a truth with specific features, that we can deal with them in particular ways. When we are not using the hammer it is a Thing in the world, present-at-hand, and much as we may look at it we cannot access its nature. When we deploy it in a task, however, we start to gain some sense of it as-it-is; we focus not on it, but on the work to be done. It is, un-intuitively, in this *un-focusing*, that we begin to encounter something of the hammer's fundamental nature, the hammer-as-hammer, rather than some theorized entity. When something is ready-to-hand we cease to concern ourselves with its nature; conscious consideration bars us from a truly embodied relationship (in Ihde's terms), from true incorporation (in mine). For Heidegger this kind of encounter leads us to come closer to the object as, at least to some extent, what it is.

4 In *Challenging the Phenomena of Technology* (Hayler 2015), I describe this melting away as being conditional upon *expert* use; it is the experienced user who no longer encounters the artefact that she deploys.

5 Ihde describes the phenomenological experience of viewing shapes and objects, such as the Necker cube, from the same vantage, but with a different perception (the Necker cube is a 2D line drawing of a 3D cube that either appears to have its front face at the level of the paper with depth pushing 'back', or its rear face at the level of the paper with its depth projecting 'out'; an image search for 'necker cube' will, of course, bring up a host of examples. With a little practice you can switch between these different 'views' of this same image, and this is what Ihde intends by a 'multistable visual effect'. See chapter four of (Ihde 2012) for a full discussion of various effects of this kind.

6 Hermeneutic phenomenology aims to interpret (rather than simply describe) the world; it recognizes the effects of culture on perception and particularly on meaning which is never direct and must always detour through cultural influence – Ihde's hermeneutic relationships take their name from their connection with this approach.

7 In a similar context, theoretical and empirical work on metaphor from, for example, Lakoff and Johnson (2003), Matlock et al. (2014) and Landau et al. (2010) further support this philosophical discussion which becomes incredibly nuanced in its specific effects in the broad field of cognitive linguistics. I discuss some of the links between Ihde's work and other research on metaphor in more detail in chapter three of *Challenging the Phenomena of Technology* (Hayler 2015).

8 A nice discussion of the history of linguistic relativism, its relationship with cognitive science and its slow return to plausibility can be found in Swoyer (2010).

9 For a collected volume of Slinkachu's work in this vein, see Slinkachu (2008).

10 The term comes from the analogue camera apparatus and lenses that allowed for perspective control. The effect is now, most often, produced digitally in post-production.

11 An image search for 'tron' will make the point instantly clear if you're unfamiliar with the film.

12 I'm indebted to Amy Cook for raising the question of whether I'm conceiving of grammar here as post hoc (as we might assume from discussion of language in cognitive linguistics); that is, does grammar function as an understanding of our reception and interpretation of what has already occurred rather than inhere in the brain or system? For the most part, yes, this is what I intend, following grammar's role in language as the underlying model. But that grammars are largely interpretative doesn't discount their role in shaping future experience and action; they also set the stage for future intention. In the examples above and to come we can better read the performers' actions with the new grammatical rules that I want to describe, indeed the performers might themselves better understand the weight of their own performance, after the fact, by considering the grammars that I argue they are drawing on. But that sense of there having been something 'in the air' which prompted the performances' particular appearance stems from a pre-, semi- or fully conscious awareness that movements which look a particular way, or things which have particular features, are likely to be interpreted in a particular fashion. Returning to language, grammar may be functionally post hoc, but I write and speak with a more or less accurate eye to how my locutions will or may be received based on prior encounters.

13 Both clips feature performances by Marquese 'Nonstop' Scott.

14 I'll leave the reader to do some searching around those terms! I can recommend the *Wikipedia* series on 'popping' as a good starting point, but, roughly, popping involves the tension and swift release of muscles; locking involves rapid movements to briefly held freezes; waving is the production of a wave of motion across the dancer's body, for example, from left-hand fingertips to right-hand fingertips; botting is better known as 'doing the robot', that is, imitating mechanical movements; and tutting is a form focused on the movements of the arms and fingers (a YouTube search for 'this is tutting' is a wonderful rabbit hole to fall down into …).

15 For a while now, I've been collecting images and things that seem to approach a new aesthetic of the future, which sounds more portentous than I mean. What I mean is that we've got frustrated with the NASA extropianism space-future, the failure of jetpacks, and we need to see the technologies we actually have with a new wonder. Consider this a mood-board for unknown products. Bridle (2011)

16 The panel included Bridle; Aaron Cope (artist, blogger and developer); Ben Terrett (designer); Joanne McNeil (writer, art critic and editor); and Russell Davies (blogger, writer and journalist). The five offered position papers on the nature of the New Aesthetic Bridle (2012); its historical corollaries McNeil (2012); its presence in contemporary visual culture Terrett (2012) and society Cope (2012); and some of its literary

implications Davies (2012). The blog posts cited in this note offer outlines of the speakers' talks (to differing levels of granularity).

17 The reader might also like to search for Eliot Elisofon's 'marcel duchamp descends staircase', a 1958 stroboscopic photograph of the painter enacting his painting.

References

actionjawa, 'The Matrix – Bullet Time + Helipad Fight Scene Super High Quality', Youtube, 26 April 2010, https://www.youtube.com/watch?v=KNrSNcaYiZg (Accessed 21 October 2014).

Aglioti, Salvatore and Mariella Pazzaglia, 'Representing Actions through Their Sound', *Experimental Brain Research*, Vol. 206 (2010), pp. 141–51.

Aldama, Frederick Luis, *Toward a Cognitive Theory of Narrative Acts*, Austin: University of Texas Press, 2010.

Andrewes, Lancelot, *The Pattern of Catechistical Doctrine at Large*, London: Roger Norton, 1650.

Ansell-Pearson, Keith, 'Book Review: Evan Thompson, *Mind in Life: Biology, Phenomenology, and the Sciences of Mind*', *Phenomenology and the Cognitive Sciences*, Vol. 8 (2009), pp. 151–8.

Anshel, Mark H., *Sport Psychology: From Theory to Practice*, 4th ed, San Francisco: Benjamin Cummings, 2003.

Armstrong, Paul B., *How Literature Plays with the Brain: The Neuroscience of Reading and Art*, Baltimore: Johns Hopkins University Press, 2013.

Babington, Gervase, *Certaine Plaine, Briefe, and Comfortable Notes Upon Euerie Chapter of Genesis*, London: for Tomas Charde, 1592.

Bachner-Melman, Rachel, Christian Dina, Ada H. Zohar, Naama Constantini, Elad Lerer, Sarah Hoch and Sarah Sella et. al., 'AVPR1a and SLC6A4 Gene Polymorphisms Are Associated with Creative Dance Performance', *PloS Genetics*, Vol. 1 (2005), pp. 394–403.

Baillargeon, Renée, Rose M. Scott and Zijing He, 'False-Belief Understanding in Infants', *Trends in Cognitive Science*, Vol. 14, No. 3 (March 2010), pp. 110–18.

Barba, Eugenio, 'Words and Presence', translated by Judy Barba, in Luis Masgrau (ed.), *Theatre: Solitude, Craft, Revolt*, Aberystwyth: Black Mountain Press, 1999, pp. 79–84.

Barba, Eugenio, 'An Amulet Made of Memory: The Significance of Exercises in the Actor's Dramaturgy', in Phillip Zarrilli (ed.), *Acting (Re)Considered*, New York: Routledge, 2002, pp. 99–105.

Barba, Eugenio and Nicola Savarese, *A Dictionary of Theatre Anthropology: The Secret Art of the Performer*, translated by Richard Fowler, London: Routledge, 1991.

Barber, Stephen, *Muybridge: The Eye in Motion*, Washington, DC: Solar, 2012.

Barker, Clive, *Theatre Games: A New Approach to Drama Training*, London: Eyre Methuen, 1997.

Barrie, James M., *The Admirable Crichton*, New York: Samuel French, 1918.

Barsalou, Lawrence W., 'Perceptual Symbol Systems', *Behavioral and Brain Sciences*, Vol. 22, No. 4 (1999), pp. 577–660.

Barsalou, Lawrence W., 'Grounded Cognition: Past, Present, and Future', *Topics in Cognitive Science,* Vol. 2, No. 4 (2010), pp. 716–24.

Barton, Bruce, 'Navigating Turbulence: The Dramaturg in Physical Theatre', *Theatre Topics,* Vol. 15, No. 1 (2005), pp. 103–19.

Bassok, Miriam and Laura Novick, 'Problem Solving', in Keith J. Holyoak and Robert G. Morrison (eds), *The Oxford Handbook of Thinking and Reasoning,* Oxford Handbooks Online, 2012, pp. 413–32.

Beilock, Sian L., *Choke: What the Secrets of the Brain Reveal About Getting It Right When You Have To,* New York: Free Press, 2010.

Beilock, Sian L., and Thomas H. Carr, 'On the Fragility of Skilled Performance: What Governs Choking under Pressure?', *Journal of Experimental Psychology: General,* Vol. 130 (2001), pp. 701–25.

Beilock, Sian L., Thomas H. Carr, Clare MacMahon and Janet L. Starkes, 'When Paying Attention Becomes Counterproductive: Impact of Divided Versus Skill-Focused Attention on Novice and Experienced Performance of Sensorimotor Skills', *Journal of Experimental Psychology: Applied,* Vol. 8 (2002), pp. 6–16.

Bergen, Benjamin, *Louder Than Words: The New Science of How the Mind Makes Meaning,* New York: Basic Books, 2012.

Bergen, Benjamin K., and Nancy Chang, 'Embodied Construction Grammar in Simulation-Based Language Understanding', in Jan-Ola Östman and Mirjam Fried (eds), *Construction Grammars: Cognitive Grounding and Theoretical Extensions,* Amsterdam: John Benjamins Publishing, 2005, pp. 147–90.

Berkovich-Ohana, A., J. Glicksohn and A. Goldstein, 'Studying the Default Mode and Its Mindfulness-Induced Changes Using EEG Functional Connectivity', *Social Cognitive and Affective Neuroscience,* Vol. 9, No. 10 (2014), pp. 1616–24.

Blair, Rhonda, 'The Method and the Computational Theory of Mind', in David Krasner (ed.), *Method Acting Reconsidered: Theory, Practice, Future,* New York: St. Martins, 2000.

Blair, Rhonda, *The Actor, Image, and Action: Acting and Cognitive Neuroscience,* New York: Routledge, 2008.

Bläsing, Bettina, Martin Puttke and Thomas Schack, *The Neurocognition of Dance: Mind, Movement and Motor Skills,* London: Psychology Press, 2010.

Bolens, Guillemette, *The Style of Gestures: Embodiment and Cognition in Literary Narrative,* Baltimore: Johns Hopkins University Press, 2012.

Booth, Michael, *Theatre in the Victorian Age,* Cambridge: Cambridge University Press, 1991.

Booth, Wayne C., *A Rhetoric of Irony,* Chicago: University of Chicago Press, 1974.

Brecht, Bertolt, 'Emphasis on Sport', translated by John Willet, in John Willet (ed.), *Brecht on Theatre: The Development of an Aesthetic,* 6th edn, New York: Hill and Wang, 1957.

Bridges, Elizabeth, 'Utopia Thought the Back Door: Kleist's Marionettes and the Mechanics of Self-Consciousness', *Seminar – A Journal of Germanic Studies,* Vol. 48, No. 1 (2012), pp. 75–90.

Bridle, James, 'About', The New Aesthetic, 6 May 2011, http://new-aesthetic. tumblr.com/about (Accessed 21 October 2014).

Bridle, James, 'The New Aesthetic', Really Interesting Group, 6 May 2011, http://www.riglondon.com/blog/2011/05/06/the-new-aesthetic (Accessed 21 October 2014).

Bridle, James, '#Sxaesthetic', booktwo.org, 15 March 2012, http://booktwo.org/ notebook/sxaesthetic (Accessed 21 October 2014).

Brincker, M., 'Navigating Beyond "Here & Now" Affordances: On Sensorimotor, 1–5. Maturation and "False Belief" Performance"', *Frontiers in Psychology*, Vol. 5, No. 1433 (2014).

Brown, Kirk Warren, Richard M. Ryan and J. David Creswell, 'Mindfulness: Theoretical Foundations and Evidence for Its Salutary Effects', *Psychological Inquiry*, Vol. 18, No. 4 (2007), pp. 211–37.

Brown, William M., Lee Cronk, Keith Grochow, Amy Jacobson, C. Karen Liu, Zoran Popovic and Robert Trivers, 'Dance Reveals Symmetry Especially in Young Men', *Nature*, Vol. 438 (2005), pp. 1148–50.

Burge, Stuart, William Shakespeare's 'Julius Caesar', USA: Commonwealth United Entertainment, 1970.

Burke, Kenneth, *A Grammar of Motives*, Berkeley: University of California Press, 1969.

Burnham, Denis, 'Language Specific Speech Perception and the Onset of Reading', *Reading and Writing*, Vol. 16 (2003), pp. 573–609.

Cahn, Rael B., and J. Polich, 'Meditation States and Traits: EEG, ERP, and Neuroimaging Studies', *Psychological Bulletin*, Vol. 132, No. 2 (2006), pp. 180–211.

Callow, Nichola and Ross Roberts, 'Imagery Research: An Investigation of Three Issues', *Psychology of Sport and Exercise*, Vol. 11, No. 5 (2010), pp. 327–29.

Capelli, Carol A., Noreen Nakagawa and Cary M. Madden, 'How Children Understand Sarcasm: The Role of Context and Intonation', *Child Development*, Vol. 61, No. 6 (1990), pp. 1824–41.

Carlson, Marla, *Performing Bodies in Pain*, London: Palgrave Macmillan, 2010.

Carney, Dana R., Amy J. C. Cuddy and Andy J. Yapp, 'Power Posing: Brief Nonverbal Displays Affect Neuroendocrine Levels and Risk Tolerance', *Psychological Science*, Vol. 21, No. 10 (2010), pp. 1363.

Chaffin, Roger and Topher R. Logan, 'Practicing Perfection: How Concert Soloists Prepare for Performance', *Advances in Cognitive Psychology*, Vol. 2, No. 2–3 (2006), pp. 113–30.

Chaffin, Roger, Topher R. Logan and Kristen T. Begosh, 'Performing from Memory', in S. Hallam, I. Cross and M. Thaut (eds), *Oxford Handbook of Music Psychology*, Oxford: Oxford University Press, 2008, pp. 352–63.

Chemero, Anthony, *Radical Embodied Cognitive Science*, Cambridge, MA: MIT University Press, 2009.

Chiel, Hillel J., and Randall D. Beer, 'The Brain Has a Body: Adaptive Behavior Emerges from Interactions of Nervous System, Body and Environment', *Trends in Neuroscience*, Vol. 20, No. 12 (1997), pp. 553–7.

Clancy, William J., 'Scientific Antecedents of Situated Cognition', in
Philip Robbins and Murat Aydede (eds), *The Cambridge Handbook of
Situated Cognition*, New York: Cambridge University Press, 2009, pp. 11–34.

Clark, Andy, *Being There*, Cambridge, MA: MIT University Press, 1997.

Clark, Andy, 'Language, Embodiment, and the Cognitive Niche', *Trends in
Cognitive Science*, Vol. 10, No. 8, (2006), pp. 370–74.

Clark, Andy, *Super-Sizing the Mind: Embodiment, Action, and Cognitive
Extension*, Oxford: Oxford University Press, 2008.

Clark, Andy and David Chalmers, 'Embodied, Situated, and Distributed
Cognition', in William Bechtel and George Graham (eds), *A Companion to
Cognitive Science*, Hoboken: Blackwell Publishing, 1998, pp. 506–17.

Clark, Andy and David Chalmers, 'Extended Mind', *Analysis*, Vol. 58, No. 1
(1998), pp. 7–19.

Clark, Herbert H., and Richard J. Gerrig, 'On the Pretense Theory of Irony',
Journal of Experimental Psychology: General, Vol. 113, No. 1 (1984),
pp. 121–6.

Cohen, Bonnie Bainbridge, *Sensing, Feeling, and Action: The Experiential
Anatomy of Body-Mind Centering*, Northampton: Contact Editions, 1993.

Cohen, Robert, *Acting Power: The 21st Century Edition*, London and New York:
Routledge, 2013.

Cook, Amy, 'Interplay: The Method and Potential of a Cognitive Scientific
Approach to Theatre', *Theatre Journal*, Vol. 59, No. 4 (2007), pp. 579–94.

Cook, Amy, *Shakespearean Neuroplay; Reinvigorating the Study of Dramatic Texts
and Performance through Cognitive Science*, London: Palgrave Macmillan, 2010.

Cook, Susan Wagner and Susan Goldin-Meadow, 'The Role of Gesture in
Learning: Do Children Use Their Hands to Change Their Minds', *Journal of
Cognition and Development*, Vol. 7, No. 2 (2006), pp. 211–32.

Cope, Aaron, 'The New Aesthetic', this_is_aaronland.info, 13 March 2012,
http://www.aaronland.info/weblog/2012/10/08/signs/#stories (Accessed 21
October 2014).

Crane, Mary Thomas, *Shakespeare's Brain: Reading with Cognitive Theory*,
Princeton and Oxford: Princeton University Press, 2001.

Csikzentmihalyi, Mihaly, *Creativity: Flow and the Psychology and Discovery of
Invention*, New York: Harper Perennial, 1997.

Cuddy, Amy, 'Your Body Language Shapes Who You Are', 2012, http://www.ted.
com/talks/amy_cuddy_your_body_language_shapes_who_you_are.

Cutrer, L. Michelle, 'Time and Tense in Narrative and in Everyday Language',
University of California, San Diego, 1994.

Damasio, Antonio, *Descartes' Error: Emotion, Reason, and the Human Brain*,
New York: Avon Books, Inc., 1994.

Damasio, Antonio, *The Feeling of What Happens: Body and Emotion in the
Making of Consciousness*, New York: Harcourt Brace & Company, 1999.

Dancygier, Barbara, 'Narrative Anchors and the Processes of Story
Construction: The Case of Margaret Atwood's *the Blind Assasin*', *Style*,
Vol. 41, No. 2 (2007), pp. 133–52.

Dancygier, Barbara, 'The Text and the Story: Levels of Blending in Fictional Narratives', in Todd Oakley and Anders Hougaard (eds), *Mental Spaces in Discourse and Interaction*, Amsterdam and Philadelphia: John Benjamins Publishing, 2008, pp. 51–78.

Dancygier, Barbara, 'Alternativity in Poetry and Drama: Textual Intersubjectivity and Framing', *English Text Construction*, Vol. 3, No. 2 (2010), pp. 165–84.

Dancygier, Barbara, *The Language of Stories: A Cognitive Approach*, Cambridge: Cambridge University Press, 2012.

Dancygier, Barbara, *Viewpoint in Language: A Multimodal Perspective*, Cambridge: Cambridge University Press, 2012.

Dancygier, Barbara, and Eve Sweetser., *Mental Spaces in Grammar: Conditional Constructions,*. Cambridge, UK: Cambridge University Press, 2005.

Dane, Erik, 'Paying Attention to Mindfulness and Its Effects on Task Performance in the Workplace', *Journal of Management*, Vol. 37, No. 4 (2011), pp. 997–1018.

Danziger, Shai, Jonathan Levav and Liora Avnaim-Pesso, 'Extraneous Factors in Judicial Decisions', *Proceedings of the National Academy of Sciences*, Vol. 108, No. 17 (2011), pp. 6889–92.

Davies, Russell, 'SXSW, the New Aesthetic and Writing', Russell Davies, 14 March 2012, http://russelldavies.typepad.com/planning/2012/03/sxsw-the-new-aesthetic-and-writing.html (Accessed 21 October 2014).

De Jaeghar, Hanne and Ezequiel Di Paolo, 'Participatory Sense-Making: An Enactive Approach to Social Cognition', *Phenomenology and the Cognitive Sciences,* Vol. 6 (2007), pp. 485–507.

Desmond, Jane C., *Meaning in Motion: New Cultural Studies of Dance*, Durham: Duke University Press, 1997.

DiBenedetto, Stephen, *The Provocation of the Senses in Contemporary Theatre*, London: Routledge, 2010.

Dissanayake, Ellen, 'Antecedents of the Temporal Arts in Early Mother-Infant Interaction', in Steven Brown, Björn Merker, Christina Wallin and Nils L. Wallin (eds), *The Origins of Music*, Cambridge, MA: The MIT Press, 2000, pp. 389–410.

Donald, Merlin, *A Mind So Rare: The Evolution of Human Consciousness*, New York: W. W. Norton, 2001.

Dreyfus, Hubert L., 'Intelligence without Representation: Merleau-Ponty's Critique of Mental Representation', *Phenomenology and the Cognitive Sciences*, Vol. 1 (2002), pp. 367–83.

Duchan, Judith, Gail Bruder and Lynn Hewitt, *Deixis in Narrative: A Cognitive Science Perspective*, Hillsdale: Lawrence Erlbaum Associates, 1995.

Edelman, Gerald, and Giulio Tononi, *A Universe of Consciousness: How Matter Becomes Imagination*, New York: Basic Books, 2000.

Emigh, John, *Masked Performance: The Play of Self and Other in Ritual and Theatre*, Philadelphia: University of Pennsylvania Press, 1996.

Emigh, John, 'Performance, Neuroscience, and the Limits of Culture', in Nathan Stuckey and Cynthia Wimmer (eds), *Teaching Performance Studies*, Southern Illinois University Press, 2002.

Emigh, John, 'Minding Bodies: Demons, Masks, Archetypes, and the Limits of Culture', *Journal of Dramatic Theory and Criticism*, Vol. 25, No. 2 (2011), pp. 125–39.

Ericsson, K. Anders, 'Deliberate Practice and the Modifiability of Body and Mind: Toward a Science of the Structure and Acquisition of Expert and Elite Performance', *International Journal of Sport Psychology*, Vol. 38, No. 1 (2007), p. 4.

Ericsson, K. Anders, and Neil Charness, 'Expert Performance: Its Structure and Acquisition', *American Psychologist*, Vol. 49, No. 8 (1994), p. 725.

Eustis, Morton, 'Paul Muni: A Profile and a Self-Portrait', in Laurence Senelick (ed.), *Theatre Arts on Acting*, London and New York: Routledge, 2008, pp. 18–24.

Evans, Jonathan, 'Spot the Difference: Distinguishing Between Two Kinds of Processing', *Mind and Society*, Vol. 11 (2012), pp. 121–32.

Farnell, Brenda, 'Moving Bodies, Acting Selves', *Annual Review of Anthropology*, Vol. 28 (1999), pp. 341–73.

Fauconnier, Gilles, *Mental Spaces: Aspects of Meaning Construction in Natural Language*, New York: Cambridge University Press, 1985.

Fauconnier, Gilles and Mark Turner, 'Compression and Global Insight', *Cognitive Linguistics*, Vol. 11, No. 3/4 (2000), pp. 283–304.

Fauconnier, Gilles and Mark Turner, *The Way We Think: Conceptual Blending and the Mind's Hidden Complexities*, New York: Basic Books, 2002.

Fillmore, Charles, 'Frames and the Semantics of Understanding', *Quaderni di Semantica*, Vol. 6 (1985), pp. 222–54.

Fillmore, Charles, 'Frame Semantics', in Dirk Geererts (ed.), *Cognitive Linguistics: Basic Readings*, Berlin: de Gruyter, 2006, pp. 373–400.

Fitzgerald, Percy, *The World Behind the Scenes*, London: Benjamin Blom, 1881.

Fivush, Robyn, 'Constructing Narrative, Emotions, and Self in Parent-Child Conversations About the Past', in Ulric Neisser and Robyn Fivush (eds), *The Remembering Self: Construction and Accuracy in the Self-Narrative*, New York: Cambridge University Press, 1994, pp. 136–57.

Fletcher, David and Mustafa Sarkar, 'A Grounded Theory of Psychological Resilience in Olympic Champions', *Psychology of Sport and Exercise*, Vol. 13, No. 5 (2012), pp. 672–74.

Fournier, Jean F., Sophie Deremaux and Marjorie Bernier, 'Content, Characteristics, and Function of Mental Imagery', *Psychology of Sport and Exercise*, Vol. 9 (2008), pp. 735–40.

Fowler, Henry Watson, *A Dictionary of Modern English Usage*, Oxford: Clarendon Press, 1926.

Frick-Horbury, Donna and Robert E. Guttentag, 'The Effects of Restricting Hand Gesture Production on Lexical Retrieval and Free Recall', *Journal of Psychology*, Vol. 111, No. 1 (1998), pp. 43–62.

Gallagher, Shaun, *How the Body Shapes the Mind*, Oxford: Oxford University Press, 2005.

Gallagher, Shaun, 'Philosophical Antecedents of Situated Cognition', in Philip Robbins and Murat Aydede (eds), *The Cambridge Handbook of Situated Cognition*, Cambridge: Cambridge University Press, 2009, pp. 35–52.

Gallagher, Shaun, 'The Socially Extended Mind', *Cognitive Systems Research*, Vol. 25/26, No. 25–26 (2013), pp. 4–12.

Gallagher, Shaun, 'Phenomenology and Embodied Cognition', in Lawrence Shapiro (ed.), *The Routledge Handbook in Embodied Cognition*, London: Routledge, 2014, pp. 9–18.

Gallagher, Shaun, Brunce Janz, Lauren Reinerman, Patricia Bockelman and Jörg Trempler, *A Neurophenomenology of Awe and Wonder: Towards a Non-Reductionist Cognitive Science*, London: Palgrave Macmillan, October 2015.

Gallagher, Shaun and Dan Zahavi, *The Phenomenological Mind: An Introduction to Philosophy of Mind and Cognitive Science*, New York: Routledge, 1998.

Gallese, Vittorio and George Lakoff, 'The Brain's Concepts: The Role of the Sensory-motor System in Conceptual Knowledge', *Cognitive Neuropsychology*, Vol. 22, No. 3/4 (2005), pp. 455–79.

Geeves, Andrew, Doris J. F. McIlwain, John Sutton and Wayne Christensen, 'To Think or Not to Think: The Apparent Paradox of Expert Skill in Music Performance', *Educational Philosophy and Theory*, Vol. 46, No. 6 (2013), pp. 674–91.

Gibbs, Raymond W., 'On the Psycholinguistics of Sarcasm', *Journal of Experimental Psychology: General*, Vol. 115, No. 1 (1986), p. 3.

Gibbs, Raymond W., *Embodiment and Cognitive Science*, Cambridge: Cambridge University Press, 2005.

Gibbs, Raymond W., 'Metaphor Interpretation as Embodied Simulation', *Mind and Language*, Vol. 21, No. 3 (2006), pp. 434–58.

Gibson, James J., *The Ecological Approach to Visual Perception*, Boston: Houghton-Mifflin, 1979.

Gill, Satinder P., 'Rhythmic Synchrony and Mediated Interaction: Towards a Framework of Rhythm in Embodied Interaction', *AI & Society*, Vol. 27 (2012), pp. 111–27.

Glanvill, Joseph, *The Vanity of Dogmatizing*, London: EC, 1661.

Glass, Renee, 'The Audience Response Tool (A.R.T.): The Impact of Choreographic Intention, Information and Dance Expertise on Psychological Reactions to Contemporary Dance', Unpublished doctoral dissertation, University of Western Sydney, 2006.

Glenberg, Arthur M., and Michael P. Kaschak, 'Grounding language in action', *Psychonomic Bulletin & Review*, Vol. 9, No. 3 (2002), pp. 558–65.

Glenberg, Arthur M., Marc Sato, Luigi Cattaneo, Lucia Riggio, Daniele Palumbo and Giovanni Buccino, 'Processing Abstract Language Modulates Motor System Activity', *Quarterly Journal of Experimental Psychology*, Vol. 61, No. 6 (2008), pp. 905–19.

Grezes Julie, Jorge Armony, James Rowe, and Richard Passingham, 'Activations Related to "Mirror" and "Canonical" Neurons in the Human Brain: An FMRI Study', *NeuroImage*, Vol. 18 (2003), pp. 928–37.

Grotowski, Jerzy, 'The Theatre's New Testament', in Eugenio Barba (ed.), *Towards a Poor Theatre*, London: Methuen, 1968, pp. 27–54.

Grove, Robin, Catherine Stevens and Shirley McKechnie, *Thinking in Four Dimensions: Creativity and Cognition in Contemporary Dance*, Carlton: Melbourne University Press, 2005.

Hacking, Ian, 'The Looping Effects of Human Kinds', in Dan Sperber, David Premack and Ann James Premack (eds), *Casual Cognition: A Multidisciplinary Approach*, New York: Oxford University Press, 1995, pp. 351–83.

Haiman, John, *Talk Is Cheap: Sarcasm, Alienation, and the Evolution of Language*, New York: Oxford University Press, 1998.

Hatzigeorgiadis, Antonis, 'Negative Self-Talk During Sport Performance: Relationships with Pre-Competition Anxiety and Goal-Performance Discrepancies', *Journal of Sport Behavior,* Vol. 31, No. 3 (2008), pp. 244–5.

Hayler, Matt, 'Incorporating Technology: A Phenomenological Approach to the Study of Artifacts and the Popular Resistance to E-Reading', PhD Thesis, University of Exeter, 2011.

Hayler, Matt, *Challenging the Phenomena of Technology*, Basingstoke: Palgrave Macmillan, 2015.

Heidegger, Martin, *Being and Time*, translated by John Macquarrie and Edward Robinson, Malden: Blackwell Publishing, 1962.

Hogarth, Robin, *Educating Intuition*, Chicago: University of Chicago Press, 2001.

Hollander, Anne, *Seeing through Clothes*, Berkeley: University of California Press, 1993.

Horan, Roy, 'The Neuropsychological Connection between Creativity and Meditation', *Creativity Research Journal*, Vol. 21, No. 2–3 (2009), pp. 199–222.

Huston, Hollis, *The Actor's Instrument: Body, Theory, Stage*, Ann Arbor: University of Michigan Press, 1992.

Hutchins, Edwin, *Cognition in the Wild*, Cambridge, MA: MIT University Press, 1995.

Hutchins, Edwin, 'Material Anchors for Conceptual Blends', *Journal of Pragmatics,* Vol. 37, No. 10 (2005), pp. 1555–77.

Hutchins, Edwin, 'The Cultural Ecosystem of Human Cognition', *Philosophical Philosophy,* Vol. 27, No. 1 (2014), pp. 34–49.

Ihde, Don, *Technology and the Lifeworld*, Bloomington: University of Indiana Press, 1990.

Ihde, Don, *Postphenomenology and Technoscience: The Peking University Lectures*, New York: State University of New York Press, 2009.

Ihde, Don, *Embodied Technics*, Copenhagen, Denmark: Automatic Press, 2010.

Ihde, Don, *Experimental Phenomenology*, Albany: State University of New York Press, 2012.

Jeannerod, Marc, *Motor Cognition: What Actions Tell the Self*, Oxford: Oxford University Press, 2006.

Jenkins, Anthony, *The Making of Victorian Drama*, Cambridge: Cambridge University Press, 1991.

Johnson, Mark, *The Body in the Mind: The Bodily Basis of Meaning, Imagination, and Reason*, Chicago: Chicago University Press, 1987.

Johnson, Mark, *The Meaning of the Body: Aesthetics of Human Understanding*, Chicago: Chicago University Press, 2007.

Kemp, Rick, *Embodied Acting: What Neuroscience Tells Us About Performance*, London: Routledge, 2012.

Keysar, Boaz, 'The Illusory Transparency of Intention: Linguistic Perspective Taking in Text', *Cognitive Psychology*, Vol. 26, No. 2 (1994), pp. 165–208.

Keysar, Boaz, 'The Illusory Transparency of Intention: Does June Understand What Mark Means Because He Means It?', *Discourse Processes*, Vol. 29, No. 2 (2000), pp. 161–72.

Kimmerle, Marliese and Paulette Côté-Laurence, *Teaching Dance Skills: A Motor Learning and Development Approach*, Andover: Michael J. Ryan, 2003.

Kirschenbaum, Daniel S., and David A. Wittrock, 'Cognitive-Behavioral Interventions in Sport: A Self-Regulatory Perspective', in John M. Silva and Robert S. Weinberg (eds), *Psychological Foundations of Sports*, Champaign: Human Kinetics Publishers, 1984, pp. 81, 90–100.

Kirschner, Sebastian and Michael Tomasello, 'Joint Drumming: Social Context Facilitates Synchronization in Preschool Children', *Journal of Experimental Child Psychology*, Vol. 102, No. 3 (2009), pp. 299–314.

Kirschner, Sebastian and Michael Tomasello, 'Joint Music Making Promotes Prosocial Behavior in 4-Year-Old Children', *Evolution and Human Behavior*, Vol. 31, No. 5 (2010), pp. 354–64.

Kirsh, David and Paul Maglio, 'On Distinguishing Epistemic from Pragmatic Action', *Cognitive Science*, Vol. 18, No. 4 (1994), pp. 513–49.

Klata, Jan, William Shakespeare's 'Hamlet', Gdańsk, Poland: Teatr Szekspirowski, 2014.

Kohler, Evelyne, Christian Keysers, M. Alessandra Umilta, Leonardo Fogassi, Vittorio Gallese and Giacomo Rizzolatti, 'Hearing Sounds, Understanding Actions: Action Representation in Mirror Neurons', *Science*, Vol. 297, No. 5582 (2002), pp. 846–8.

Kozak, Megan N., Tomi-Ann Roberts and Kelsey E. Patterson, 'She Stoops to Conquer? How Posture Interacts with Self-Objectification and Status to Impact Women's Affect and Performance', *Psychology of Women Quarterly*, Vol. 38, No. 3 (2014), p. 415.

Kreuz, Roger J., and Sam Glucksberg, 'How to Be Sarcastic: The Echoic Reminder Theory of Verbal Irony', *Journal of Experimental Psychology: General*, Vol. 118, No. 4 (1989), p. 374.

Lakoff, George and Mark Johnson, *Philosophy in the Flesh*, New York: Basic Books, 1999.

Lakoff, George and Mark Johnson, *Metaphors We Live By*, Chicago: University of Chicago Press, 2003.

Landau, Ayelet N., Lisa Aziz-Zadeh and Richard B. Ivry, 'The Influence of Language on Perception: Listening to Sentences About Faces Affects the Perception of Faces', *Journal Neuroscience*, Vol. 30 (2010), pp. 15254–61.

Langacker, Ronald W., '*Universals of Construal*', In *Proceedings of the Nineteenth Annual Meeting of the Berkeley Linguistics Society: General Session and Parasession on Semantic Typology and Semantic Universals* (1993), pp. 447–63.

Langacker, Ronald W., *Cognitive Grammar: A Basic Introduction*, Oxford: Oxford University Press, 2008.

Langer, Ellen, *The Power of Mindful Learning*, Boston, ON: Addison-Wesley Publishing Co., 1997.

Langer, Ellen and Minhea Moldoveanu, 'The Construct of Mindfulness', *Journal of Social Issues*, Vol. 56, No. 1 (2000), pp. 1–9.

LaPiere, Richard T. 'Attitudes Versus Actions'. *Social Forces*, Vol. 3, No. 2 (1934), pp. 230–37.

Lave, Jean, 'Situating Learning in Communities of Practice', *Perspectives on Socially Shared Cognition*, Vol. 2 (1991), pp. 63–82.

Leman, Marc, Frank Desmet, Frederik Styns, Leon Van Noorden and Dirk Moelants, 'Sharing Musical Expression through Embodied Listening: A Case Study Based on Chinese Guqin Music', *Music Perception*, Vol. 26, No. 3 (2009), pp. 263–78.

Levesque, Chantal and Kirk Brown, 'Mindfulness as a Moderator of the Effect of Implicit Motivational Self-Concept on Day-to-Day Behavioral Motivation', *Motivation and Emotion*, Vol. 31 (2007), pp. 284–99.

Libby, Lisa K., Thomas Gilovich and Richard P. Eibach, 'Here's Looking at Me: The Effect of Memory Perspective on Assessments of Personal Change', *Journal of Personality and Social Psychology*, Vol. 88, No. 1 (2005), pp. 51–3.

Limb, Charles and Allen Braun, 'Neural Substrates of Spontaneous Musical Performance: An FMRI Study of Jazz Improvisation', *PLoS ONE*, Vol. 3, No. 2 (2008), pp. 1–9.

Lingis, Alphonso, 'The Body Postured and Dissolute', in Véronique Fóti (ed.), *Merleau Ponty: Difference, Materiality, Painting*, Atlantic Highlands, NJ: Humanities Press, 1996, pp. 60–71.

Littman, David C., and Jacob L. Mey, 'The Nature of Irony: Toward a Computational Model of Irony', *Journal of Pragmatics*, Vol. 15, No. 2 (1991), pp. 131–51.

Lizardo, Omar, 'Is a "Special Psychology" of Practice Possible? From Values and Attitudes to Embodied Dispositions', *Theory and Psychology*, Vol. 19, No. 6 (2009), pp. 713–27.

Lloyd, Phyllida, 'Julius Caesar: Resources', 2012, http://www.donmarwarehouse.com/~/media/Files/Julius%20Caesar%20Behind%20the%20Scenes%20Guide.ashx.

Lloyd, Phyllida, William Shakespeare's 'Julius Caesar', London: Donmar Warehouse, 2013.

Lupyan, Gary and Emily Ward, 'Language Can Boost Otherwise Unseen Objects into Visual Awareness', *Proceedings of the National Academy of Science of the United States of America*, Vol. 110 (2013), pp. 14196–201.

Lutterbie, John, 'Neuroscience and Creativity in the Rehearsal Process', in Bruce McConachie and F. Elizabeth Hart (eds), *Performance and Cognition: Theatre Studies and the Cognitive Turn*, New York: Routledge, 2006, pp. 149–66.

Lutterbie, John, *Toward a General Theory of Acting: Cognitive Science and Performance*, New York: Palgrave Macmillan, 2011.

MacWhinney, Brian, 'The Emergence of Grammar from Perspective', in Diane Pecher and Rolf A. Zwaan (eds), *Grounding Cognition: The Role of Perception and Action in Memory, Language, and Thinking*, Cambridge: Cambridge University Press, 2005, pp. 198–223.

Malabou, Catherine, *The Future of Hegel: Plasticity, Temporality, and Dialectic*, translated by Lisabeth During, London: Routledge, 2005.

Malabou, Catherine, *What Should We Do with Our Brain?*, translated by Sebastian Rand, New York: Fordham University Press, 2008.

Malloch, Stephen and Colwyn Trevarthen, *Communicative Musicality: Exploring the Basis of Human Companionship*, Oxford: Oxford University Press, 2009.

Mandler, Jean M. and Cristobal Pagan Cánovas, 'On Defining Image Schemas', in *Language and Cognition* (2014), pp. 1–23.

Matlock, Teenie, Spencer C. Castro, Morgan Fleming, Timothy M. Gann and Paul P. Maglio, 'Spatial Metaphors of Web Use', *Spatial Cognition & Computation*, Vol. 14, No. 4 (2014), pp. 306–20.

McConachie, Bruce, 'Doing Things with Image Schemas: The Cognitive Turn in Theatre Studies and the Problem of Experience for Historians', *Theatre Journal*, Vol. 53, No. 4 (2001), pp. 569–94.

McConachie, Bruce, *American Theater in the Culture of the Cold War*, Iowa City: University of Iowa Press, 2003.

McConachie, Bruce, *Engaging Audiences: A Cognitive Approach to Spectating in the Theatre*, New York, Basingstoke and London: Palgrave Macmillan, 2008.

McConachie, Bruce and F. Elizabeth Hart, *Performance and Cognition: Performance Studies and the Cognitive Turn*, London: Routledge, 2007.

McKechnie, Shirley, 'Dancing Memes, Minds and Designs', in Robin Grove, Catherine Stevens and Shirley McKechnie (eds), *Thinking in Four Dimensions: Creativity and Cognition in Contemporary Dance*, Carlton: Melbourne University Press, 2005, pp. 81–94.

McNeil, Joanne, 'New Aesthetic at SXSW', Joanne McNeil, 14 March 2012, http://www.joannemcneil.com/new-aesthetic-at-sxsw (Accessed 21 October 2014).

McNeill, William H., *Keeping Together in Time: Dance and Drill in Human History*,. Cambridge, MA: Harvard University Press, 1997.

Meisner, Sanford and Dennis Longwell, *Sanford Meisner on Acting*, New York: Vintage Books, 1987.

Mellor, Anne K., *English Romantic Irony*, Cambridge, MA: Harvard University Press, 1980.

Meltzoff, Andrew and M. Keith Moore, 'Imitation of Facial and Manual Expressions by Human Neonates', *Science* 198, pp. 75–77.

Meltzoff, Andrew and M. Keith Moore, 'Newborn Infants Imitate Adult Facial Gestures', *Child Dev.* Vol. 54 (1983), pp. 702–9.

Menary, Richard and Michael Kirchhoff, 'Cognitive Transformations and Extended Expertise', *Educational Philosophy and Theory*, Vol. 46, No. 6 (2013), pp. 610–23.

Merchant, Hugo, Jessica Grahn, Laurel Trainor, Martin Rohrmeier and W. Tecumseh Fitch, 'Finding the Beat: A Neural Perspective across Humans and Non-Human Primates', *Philosophical Transactions of the Royal Society of London, Series B: Biological Sciences*, Vol. 370 (2015), pp. 893–903.

Miller, Kai J., Gerwin Schalk, Eberhard E. Fetz, Marcel den Nijs, Jeffrey G. Ojemann, Rajesh P. N. Rao and Riitta Hari, 'Cortical Activity During Motor Execution, Motor Imagery, and Imagery-Based Online Feedback', *Proceedings of the National Academy of Sciences*, Vol. 107, No. 9 (2010), pp. 4430–31.

Modugno, Nicola, Sara Iaconelli, Mariagrazia Fiorilli, Francesco Lena, Imogen Kusch and Giovanni Mirabella, 'Active Theater as a Complementary Therapy for Parkinson's Disease Rehabilitation: A Pilot Study', *The Scientific World*, Vol. 10 (2010), pp. 2301–13.

Moran, Nikki, 'Social Implications Arise in Embodied Music Cognition Research Which Can Counter Musicological "Individualism"', *Frontiers in Psychology*, Vol. 5 (2014), pp. 676.

Muecke, Douglas, *The Compass of Irony*, Oxford: Oxford University Press, 1969.

Neal, David T., and Tanya L. Chartrand, 'Embodied Emotion Perception: Amplifying and Dampening Facial Feedback Modulates Emotion Perception Accuracy', *Social Psychological and Personality Science*, Vol. 2, No. 6 (2011), pp. 673–8.

Nellhaus, Tobin, *Theatre, Communication, Critical Realism (What Is Theatre?)*, London: Palgrave Macmillan, 2010.

Nieuwland, Mante S., and Jos J. A. Van Berkum, 'When Peanuts Fall in Love: N400 Evidence for the Power of Discourse', *Journal of Cognitive Neuroscience*, Vol. 18, No. 7 (2006), pp. 1098–1111.

Noë, Alva, *Action in Perception*, Cambridge, MA: MIT University Press, 2004.

Noice, Helga and Tony Noice, 'What Studies of Actors and Acting Can Tell Us About Memory and Cognitive Functioning', *Current Directions in Psychological Science*, Vol. 15, No. 1 (2006), pp. 14–18.

North, Thomas, *Plutarch's Lives of the Noble Grecians and Romans*, London: Richard Field, 1595.

O'Regan, J. Kevin, Erik Myin and Alva Noë, 'Skill, Corporeality and Alerting Capacity in an Account of Sensory Consciousness', *Progress in Brain Research*, Vol. 150 (2005), pp. 55–68.

O'Regan, J. Kevin and Alva Noë, 'A Sensorimotor Account of Vision and Visual Consciousness', *Behavioral and Brain Sciences*, Vol. 24 (2001), pp. 939–1031.

Odin Teatret Film, 'Physical Training at Odin Teatret 1972', Denmark: Odin Teatret Film, 1972, https://www.youtube.com/watch?v=s8dj4awiMJY.

Opacic, Tajana, Catherine Stevens and Barbara Tillmann, 'Unspoken Knowledge: Implicit Learning of Structured Human Dance Movement', *Journal of Experimental Psychology: Learning, Memory and Cognition*, Vol. 35, No. 6 (2009), pp. 1570–77.

Paavolainen, Teemu, *Theatre/Ecology/Cognition: Theorizing Performer-Object Interaction in Grotowski, Kantor, and Meyerhold*, London: Palgrave Macmillan, 2012.

Parry, Idris, 'Kleist on Puppets', in Idris Parry (ed.), *Hand to Mouth, and Other Essays*, Manchester: Carcanet Press, 1981, pp. 9–12.

Pfister, Manfred, *The Theory and Analysis of Drama*, New York: Cambridge University Press, 1988.

Phelps, Elizabeth A., 'Emotion and Cognition: Insights from Studies of the Human Amygdala', *Annual Review of Psychology*, Vol. 57 (2006), pp. 27–53.

Phillips-Silver, Jessica and Peter E. Keller, 'Searching for Roots of Entrainment and Joint Action in Early Musical Interactions', *Frontiers in Human Neuroscience*, Vol. 6 (2012), p. 26.

Port, Robert F., and Timothy Van Gelder, *Mind as Motion: Explorations in the Dynamics of Cognition*, Cambridge, MA: MIT University Press, 1995.

Powell, Kerry, *The Cambridge Companion to Victorian and Edwardian Theatre*, Cambridge: Cambridge University Press, 2004.

Press, Carol M., *The Dancing Self: Creativity, Modern Dance, Self Psychology, Transformative Education*, Cresskill: Hampton Press, 2002.

Prinz, Jesse J., *Furnishing the Mind: Concepts and Their Perceptual Basis*, Cambridge, MA: MIT University Press, 2002.

Prinz, Jesse J., *Gut Reactions: A Perceptual Theory of Emotion*, New York: Oxford University Press, 2004.

Prinz, Jesse J., 'Is Consciousness Embodied?', in Philip Robbins and Murat Aydede (eds), *The Cambridge Handbook of Situated Cognition*, New York: Cambridge University Press, 2009, pp. 419–36.

Recchia, Holly E., Nina Howe, Hildy S. Ross and Stephanie Alexander, 'Children's Understanding and Production of Verbal Irony in Family Conversations', *British Journal of Developmental Psychology*, Vol. 28, No. 2 (2010), p. 255.

Renaud, Lissa Tyler, 'Training Artists or Consumers? Commentary on American Actor Training', in Ellen Margolis and Lissa Tyler Renaud (eds), *The Politics of American Actor Training*, New York: Routledge, 2010, pp. 76–93.

Ricciardi, Emiliano, Daniela Bonino, Lorenzo Sani, Tomaso Vecchi, Mario Guazzelli, James V. Haxby, Luciano Fadiga and Pietro Pietrini, 'Do We Really Need Vision? How Blind People "See" the Actions of Others', *Journal of Neuroscience*, Vol. 29 (2009), pp. 9719–24.

Richardson, Alan, *British Romanticism and the Science of the Mind*, New York: Cambridge University Press, 2001.

Richardson, Alan, '"But What Then Am I, This Inexhaustible, Unfathomable, Historical Self?"', *SYNTHESE*, Vol. 178 (2010), pp. 143–54.

Riloff, Ellen, Ashequl Qadir, Prafulla Surve, Lalindra De Silva, Nathan Gilbert and Ruihong Huang, 'Sarcasm as Contrast between a Positive Sentiment and Negative Situation', In *Proceedings of the 2013 Conference on Empirical Methods in Natural Language Processing*, 2013, pp. 704–14.

Robbins, Philip and Murat Aydede, *The Cambridge Handbook of Situated Cognition*, New York: Cambridge University Press, 2009.

Rokotnitz, Naomi, *Trusting Performance: A Cognitive Approach to Embodiment in Drama*, London: Palgrave Macmillan, 2011.

Rowlands, Mark, 'Situated Representations', in Philip Robbins and Murat Aydede (eds), *The Cambridge Handbook of Situated Cognition*, Cambridge: Cambridge University Press, 2009, pp. 117–33.

Rubba, Jo, 'Alternative Ground in the Interpretation of Deictic Expressions', in Gilles Fauconnier and Eve Sweetser (eds), *Spaces, Worlds, and Grammar*, Chicago: University of Chicago Press, 1996, pp. 227–61.

Rushgaroth, 'Amazing Tron Dance Performed by Wrecking Orchestra [Better Quality]', Youtube, 16 March 2012, https://www.youtube.com/watch?v=-Rot9uaVO8s (Accessed 21 October 2014).

Rylance, Mark, 'Assassination of Caesar: Fight Sequence', In *Julius Caesar*, London: Globe Theatre Archive, 1999a.

Rylance, Mark, William Shakespeare's 'Julius Caesar', London: The Globe Theatre, 1999b.

Rylance, Mark, 'Prompt Book', In *Julius Caesar*, London: Globe Theatre Archive, 1999.

Sawyer, R. Keith, *Explaining Creativity: The Science of Human Innovation*, New York: Oxford University Press, 2006.

Schneider, Rebecca, *Theatre and History*, London: Palgrave Macmillan, 2014.

Seddon, Frederick, 'Modes of Communication During Jazz Improvisation', *British Journal of Music Education*, Vol. 22, No. 1 (2005), pp. 47–61.

Sedgewick, Garnett Gladwin, *Of Irony: Especially in Drama*, Reprint ed. Vancouver: Ronsdale Press, [1935] 2003.

Shapiro, Lawrence, *Embodied Cognition*, London: Routledge, 2010.

Shaughnessy, Nicola, *Applying Performance: Live Art, Socially Engaged Theatre, and Affective Practice*, Basingstoke: Palgrave Macmillan, 2012.

Shaughnessy, Nicola, *Affective Performance and Cognitive Science: Body, Brain and Being*, London: Methuen, 2013.

Shaughnessy, Nicola, 'Imagining Otherwise: Autism, Neuroaesthetics and Contemporary Performance', *Interdisciplinary Science Reviews*, Vol. 38, No. 4 (2013), pp. 321–34.

Slaby, Jan and Shaun Gallagher, 'Critical Neuroscience and Socially Extended Minds', *Theory, Culture & Society*, Vol. 32, No. 1 (2014), pp. 33–59.

Slinkachu, *Little People in the City: The Street Art of Slinkachu*, London: Boxtree, 2008.

Sofer, Andrew, *The Stage Life of Props*, Ann Arbor: University of Michigan Press, 2003.

Sontag, Susan, 'Notes on Camp', in Susan Sontag (ed.), *Against Interpretation and Other Essays*, New York: Farrar, Straus, and Giroux, 1964, pp. 275–92.

Sontag, Susan, *On Photography*, New York: Farrar, Straus & Giroux, 1977.

Sperber, Dan and Deirdre Wilson, 'Irony and the Use-Mention Distinction', *Radical Pragmatics*, Vol. 49 (1981), pp. 295–318.

Sperber, Dan and Deirdre Wilson, 'Irony and Relevance: A Reply to Seto, Hamamoto and Yamanashi', in Robyn Carston and Seiji Uchida (eds), *Relevance Theory: Applications and Implications*, Amsterdam: John Benjamins Publishing, 1998, pp. 283–93.

Spolsky, Ellen, *Gaps in Nature: Literary Interpretation and the Modular Mind*, Albany: State University of New York Press, 1993.

Spolsky, Ellen, *Contracts of Fiction: Cognition, Culture, Community*, Oxford: Oxford University Press, 2015.

Stambulova, Natalia, Alexander Stambulova and Urban Johnson, 'Believe in Yourself, Channel Energy, and Play Your Trumps Olympic Preparation in Complex Coordination Sports', *Psychology of Sport and Exercise*, Vol. 13, No. 5 (2012), pp. 679–86.

Stanislavski, Konstantin, *An Actor's Work: A Student's Diary*, translated by Jean Benedetti, London: Routledge, 2010.

Stanovich, Keith, *The Robot's Rebellion: Finding Meaning in the Age of Darwin*, Chicago: University of Chicago Press, 2004.

Stanovich, Keith, Richard West and Maggie Toplak, 'Individual Differences as Essential Components of Heuristics and Biases Research', in Ken Manktelow, David Over and Shira Elqayam (eds), *The Science of Reason: A Festschrift for Jonathan St. B.T Evans*, New York: Psychology Press, 2011, pp. 355–96.

Steele, Valerie, *Fashion and Eroticism: Ideals of Feminine Beauty from the Victorian Era to the Jazz Age*, Oxford: Oxford University Press, 1985.

Steenburgh, J. Jason Van, Jessica I. Fleck, Mark Beeman and John Kounios, 'Insight', in Keith J. Holyoke and Robert G. Morrison (eds), *The Oxford Handbook of Thinking and Reasoning*, New York: Oxford University Press, 2012, pp. 475–91.

Sterelny, Kim, *The Evolved Apprentice: How Evolution Made Humans Unique*, Cambridge, MA: MIT University Press, 2012.

Sterling, Bruce, 'An Essay on the New Aesthetic', *Wired*, 2 April 2012, http://www.wired.com/2012/04/an-essay-on-the-new-aesthetic (Accessed 21 October 2014).

Stern, Tiffany, 'Epilogues, Prayers after Plays, and Shakespeare's *2 Henry IV*', *Theatre Notebook*, Vol. 64 (2010), pp. 22–9.

Stevens, Catherine and Shirley McKechnie, 'Thinking in Action: Thought Made Visible in Contemporary Dance', *Cognitive Processing*, Vol. 6, No. 4 (2005a), pp. 243–52.

Stevens, Catherine and Shirley McKechnie, 'Minds and Motion: Dynamical Systems in Choreography, Creativity, and Dance', in J. Birringer and J. Fenger (eds), *Tanz Im Kopf: Yearbook 15 of the German Dance Research Society, 2004*, Münster: LIT Verlag, 2005b, pp. 241–52.

Stevens, Catherine, Jane Ginsborg and Garry Lester, 'Backwards and Forwards in Space and Time: Recalling Dance Movement from Long-Term Memory', *Memory Studies*, Vol. 4 (2011), pp. 234–50.

Stevens, Catherine, Stephen Malloch, Shirley McKechie and Nicole Steven, 'Choreographic Cognition: The Time-Course and Phenomenology of Creating a Dance', *Pragmatics & Cognition*, Vol. 11, No. 2 (2003), pp. 297–326.

Stevens, Catherine, Barbara Tillmann, Tajana Opacic, 'Unspoken Knowledge: Implicit Learning of Structured Human Dance Movement', *Journal of Experimental Psychology: Learning, Memory and Cognition*, Vol. 35, No. 6 (2009), pp. 1570–77.

Stevens, Catherine, Heather Winskel, Clare Howell, Lyne-Marine Vidal, Cyril Latimer and Josephine Milne-Home, 'Perceiving Dance: Schematic Expectations Guide Experts' Scanning of a Contemporary Dance Film', *Journal of Dance Medicine & Science*, Vol. 14, No. 1 (2010), pp. 19–25.

Stevens, Catherine and J. Leach, (2015). 'Bodystorming: Effects of Collaboration and Familiarity on Improvising Contemporary Dance', *Cognitive Processing – International Quarterly of Cognitive Science*, 16, Suppl 1 (2015), S403–S407.

Stevens, Catherine, Roger Dean, Kim Vincs, and Emery Schubert, 'In the Heat of the Moment: Audience Real Time Response to Live Music and Dance Performance', in Karen Burland and Stephanie Pitts (eds), *Coughing and Clapping: Investigating Audience Experience*, Burlington: Ashgate Publishing Company, 2014.

Stevenson, Jill, *Performance, Cognitive Theory, and Devotional Culture: Sensual Piety in Late Medieval York*, London: Palgrave Macmillan, 2010.

Stockwell, Peter, *Cognitive Poetics: An Introduction*, London: Taylor & Francis (Psychology Press), 2002.

Stockwell, Peter, *Texture – A Cognitive Aesthetics of Reading*, Edinburgh: Edinburgh University Press, 2012.

Stoppard, Tom, *Arcadia*, London: Faber and Faber, Inc., 1993.

Sutton, John, 'Batting, Habit and Memory: The Embodied Mind and the Nature of Skill', *Sport in Society*, Vol. 10, No. 5 (2007), pp. 763–86.

Sutton, John, Celia B. Harris, Paul G. Keil and Amanda J. Barnier, 'The Psychology of Memory, Extended Cognition, and Socially Distributed Remembering', *Phenomenology and the Cognitive Sciences*, Vol. 9, No. 4 (2010), pp. 521–60.

Sutton, John, Doris McIlwain, Wayne Christensen and Andrew Geeves, 'Applying Intelligence to the Reflexes: Embodied Skills and Habits between

Dreyfus and Descartes', *Journal of the British Society for Phenomenology*, Vol. 42, No. 1 (2011), pp. 78–103.

Sweetser, Eve, '"The Suburbs of Your Good Pleasure": Cognition, Culture, and the Bases of Metaphoric Structure', in Graham Bradshaw, Tom Bishop and Mark Turner (eds), *The Shakespeare International Yearbook 4*, Aldershot and Burlington: Ashgate, 2004, pp. 24–55.

Swoyer, Chris, 'How Does Language Affect Thought', University of Oklahoma, 14 April 2010, http://www.ou.edu/ouphil/faculty/swoyer/LanguageThought.pdf (Accessed 30 September 2014).

Talmy, Leonard, 'Fictive Motion in Language and Ception', in Paul Bloom (ed.), *Language and Space*, Cambridge, MA: MIT University Press, 1996, pp. 211–76.

Talmy, Leonard, *Toward a Cognitive Semantics, Volume 2*, Cambridge, MA: MIT University Press, 2000.

Targoff, Ramie, *Common Prayer: The Language of Public Devotion in Early Modern England*, Chicago: Chicago University Press, 2001.

Taylor, George, *Players and Performances in the Victorian Theatre*, Manchester: Manchester University Press, 1989.

Tepperman, Joseph, David R. Traum and Shrikanth Narayanan, '"Yeah Right": Sarcasm Recognition for Spoken Dialogue Systems', In *Proceedings of Interspeech*, 2006, pp. 1838–41.

Terrett, Ben, 'SXSW, the New Aesthetic and Commercial Visual Culture', Noisy Decent Graphics, 14 March 2012, http://noisydecentgraphics.typepad.com/design/2012/03/sxsw-the-new-aesthetic-and-commercial-visual-culture.html (Accessed 21 October 2014).

Thelen, Esther, 'Time-Scale Dynamics and the Development of an Embodied Cognition', in Robert F. Port and Timothy Van Gelder (eds), *Mind as Motion: Explorations of the Dynamics of Cognition*, Cambridge, MA: The MIT Press, 1995, pp. 69–100.

Theodorakis, Yannis, Antonis Hatzigeorgiadis and Stiliani Chroni, 'Self-Talk: It Works, but How? Development and Preliminary Validation of the Functions of Self-Talk Questionnaire', *Measurement in Physical Education & Exercise Science*, Vol. 12, No. 1 (2008), pp. 11–13.

Thompson, Evan, *Mind in Life: Biology, Phenomenology, and the Sciences of Mind*, Cambridge, MA: Belknap Press of Harvard University Press, 2007.

Tobin, Vera and Michael Israel, 'Irony as a Viewpoint Phenomenon', in Barbara Dancygier and Eve Sweetser (eds), *Viewpoint in Language: A Multimodal Perspective*, Cambridge: Cambridge University Press, 2012, pp. 25–46.

Trewin, John. C., *The Edwardian Theatre.*, Oxford: Blackwell Publishers, 1978.

Trexler, Richard C., *The Christian at Prayer: An Illustrated Prayer Manual Attributed to Peter the Chanter*, New York: Medieval and Renaissance Texts and Studies, 1987.

Tribble, Evelyn B., 'Distributing Cognition in the Globe', *Shakespeare Quarterly*, Vol. 56, No. 2 (2005), pp. 135–55.

Tribble, Evelyn B., *Cognition in the Globe: Attention and Memory in Shakespeare's Theatre*, London: Palgrave Macmillan, 2011.

Tribble, Evelyn B., and Nicholas Keene, *Cognitive Ecologies and the History of Remembering*, London: Palgrave Macmillan, 2011.

Tribble, Evelyn B., and John Sutton, 'Cognitive ecology as a framework for Shakespearean studies', *Shakespeare Studies*, Vol. 39 (2011), pp. 94–103.

Tsur, Reuven, *Poetic Rhythm: An Empirical Study in Cognitive Poetics*, Berne: Peter Lang, 2012.

Turner, Cathy and Synne Behrndt, *Dramaturgy and Performance*, New York: Palgrave Macmillan, 2008.

Ullman, Michael T., 'Contributions of Memory Circuits to Language: The Declarative/Procedural Model', *Cognition*, Vol. 92 (2004), pp. 231–70.

Utsumi, Akira, 'A Unified Theory of Irony and Its Computational Formalization', in *Proceedings of the 16th Conference on Computational Linguistics – Volume 2*, Stroudsburg: Association for Computational Linguistics, 1996, pp. 962–7.

Utterback, Neal, *Stagehands: Gestures and the Embodied Actor*, Bloomington: Indiana University, 2013.

Van Hoek, Karen, *Anaphora and Conceptual Structure*, Chicago: University of Chicago Press, 1997.

Varela, Francisco J., Eleanor Rosch and Evan Thompson, *The Embodied Mind: Cognitive Science and Human Experience*, Cambridge, MA: MIT University Press, 1991.

Verbeek, Peter-Paul, *What Things Do*, University Park: University of Pennsylvania Press, 2005.

Vervaeke, John, 'Chi Explained without Magic', Paper presented at the Performing Philosophy Symposium, University of Toronto, August 15 2011, https://www.youtube.com/watch?v=vcp6J1T60qc.

Vervaeke, John and Arianne Herrera-Bennett, 'The Insight-Cascade: A Comprehensive Conceptualization of Flow Based in the Dynamical Systems Theory of Cognition', Author draft, University of Toronto, 2013.

Vicary, Staci A., Rachel A. Robbins, Beatriz Calvo-Merino and Catherine J. Stevens, 'Recognition of Dance-Like Actions: Memory for Static Postures or Dynamic Movements?', *Memory & Cognition*, Vol. 42, No. 5 (2014), pp. 755–67.

Von Kleist, Henrich, 'On the Marionette Theatre', in Michael J. Sidnell (ed.), *Sources of Dramatic Theory*, New York: Cambridge University Press, [1810] 1991, pp. 234–40.

Wachowski, Andy and Lana Wachowski, *The Matrix*. USA: Warner Bros Entertainment, Inc., 1999.

Wacquant, Loïc, 'The Social Logic of Sparring: On the Body as Practical Strategist', in Jennifer Hargreaves and Patricia Vertinsky (eds), *Culture, Power and the Body*, London: Routledge, 2007, pp. 142–57.

Wang, Qi, 'Culture Effects on Adults' Earliest Childhood Recollection and Self-Description: Implications for the Relation between Memory and the Self', *Journal of Personality and Social Psychology*, Vol. 81, No. 2 (2001), pp. 220–33.

Warburton, Edward C., 'Knowing What It Takes: The Effect of Perceived Learner Advantages on Dance Teachers' Use of Critical-Thinking Activities', *Research in Dance Education*, Vol. 5, No. 1 (2004), pp. 69–82.

Warburton, Edward C., 'Of Meanings and Movements: Re-Languaging Embodiment in Dance Phenomenology and Cognition', *Dance Research Journal*, Vol. 43, No. 2 (2011), pp. 65–84.

Warburton, Edward C., Margaret Wilson, Molly Lynch, and Shannon Cuykendall, 'The Cognitive Benefits of Movement Reduction: Evidence from Dance Marking', *Psychological Science*, Vol. 24, No. 9 (2013), pp. 1732–39.

Weinberg, Robert S., 'Mental Preparation Strategies', in John M. Silva and Robert S. Weinberg (eds), *Psychological Foundations of Sports*, Champaign: Human Kinetics Publishers, 1984, p. 146.

West, Lindy, 'A Complete Guide to "Hipster Racism"', *Jezebel.com*, 2012, http://jezebel.com/5905291/a-complete-guide-to-hipster-racism.

WHZGUD, 'Pumped up Kicks | Dubstep', Youtube, 23 September 2011, https://www.youtube.com/watch?v=LXO-jKksQkM (Accessed 21 October 2014).

WHZGUD, 'Take Me Away | Dubstep', Youtube, 23 January 2012, ttps://www.youtube.com/watch?v=4lOUJ2XqdTw (Accessed 21 October 2014).

Wilson, Deirdre, 'The Pragmatics of Verbal Irony: Echo or Pretense?', *Lingua*, Vol. 116, No. 10 (2006), pp. 1722–43.

Wilson, Margaret, 'Six Views of Embodied Cognition', *Psychonomic Bulletin & Review*, Vol. 9, No. 4 (2002), pp. 625–36.

Wimmer, Heinz and Josef Perner, 'Beliefs About Beliefs: Representation and Constraining Function of Wrong Beliefs in Young Children's Understanding of Deception', *Cognition*, Vol. 13, No. 1 (1983), pp. 103–28.

Woodward, James, 'Causation and Manipulability', in Edward N. Zalta (ed.), *The Stanford Encyclopedia of Philosophy* (Winter 2013 Edition), http://plato.stanford.edu/archives/win2013/entries/causation-mani/.

Wystrach, Antoine, "We've Been Looking at Ant Intelligence the Wrong Way" *Scientific American*. 30 August 2013. Accessed 9 July 2015. http://www.scientificamerican.com/article/weve-been-looking-at-ant-intelligence-the-wrong-way/.

Zarrilli, Phillip, *Psychophysical Acting: An Intercultural Approach after Stanislavski*, London: Routledge, 2008.

Zohar, Anat, Adi Degani and Einav Vaaknin, 'Teachers' Beliefs About Low-Achieving Students and Higher-Order Thinking', *Teaching and Teacher Education*, Vol. 17, No. 4 (2001), pp. 469–85.

Zunshine, Lisa, *Strange Concepts and the Stories They Make Possible: Cognition, Culture, Narrative*, Baltimore: Johns Hopkins University Press, 2008.

Zwaan, Rolf A., and Lawrence J. Taylor, 'Seeing, Acting, Understanding: Motor Resonance in Language Comprehension', *Journal of Experimental Psychology*, Vol. 135, No. 1 (2006), pp. 1–11.

Index